Project Management in the Oil and Gas Industry

Scrivener Publishing
100 Cummings Center, Suite 541J
Beverly, MA 01915-6106

Publishers at Scrivener
Martin Scrivener (martin@scrivenerpublishing.com)
Phillip Carmical (pcarmical@scrivenerpublishing.com)

Project Management in the Oil and Gas Industry

Mohamed A. El-Reedy

Scrivener
Publishing

WILEY

Co-published by John Wiley & Sons, Inc. Hoboken, New Jersey, and Scrivener Publishing LLC, Salem, Massachusetts.
Published simultaneously in Canada.

For general information on our other products and services or for technical support, please contact our Customer Care Department within the United States at (800) 762-2974, outside the United States at (317) 572-3993 or fax (317) 572-4002.

Wiley also publishes its books in a variety of electronic formats. Some content that appears in print may not be available in electronic formats. For more information about Wiley products, visit our web site at www.wiley.com.

For more information about Scrivener products please visit www.scrivenerpublishing.com.

Cover design by Kris Hackerott

Library of Congress Cataloging-in-Publication Data:

ISBN 978-1-119-08361-0

Printed in the United States of America

10   9   8   7   6   5   4   3   2   1

This book is dedicated to the spirits of my mother and my father, and to my wife and my children Maey, Hisham, and Mayar.

Contents

Preface

Oil and gas projects have special characteristics that need a different technique in project management. The development of any country depends on the development of the energy reserve through investing in oil and gas projects through onshore and offshore exploration, drilling, and increasing facility capacities.

Therefore, these projects need a sort of management match with their characteristics. Project management is the main tool to achieving a successful project. This book focuses on using practical tools and methods that are widely and successfully used in project management for oil and gas projects.

Most engineers study all subjects, but focus on project management in housing projects, administration projects, and commercial buildings or other similar projects. However, oil and gas projects have their own requirements and characteristics in management from the owners, engineering offices, and contractors' side.

This book is not only addressed to graduating engineers who wish to improve their skills in project management, but also to upper level management. This book covers all of the project management subjects from an industrial point of view specifically for petroleum projects.

The aim of this book is to be helpful to any engineering discipline or staff in sharing or applying the work of a petroleum project.

Why do the senior managers accept this project and refuse the others? Why does this company have huge investments in this country, but little investments on another countries, especially in oil and gas projects? These questions are always on our mind, so it is important to discuss them clearly.

Knowledge is power. When you understand well, you will do well. Therefore, in this book, we will illustrate exactly what the project manager is thinking when they are working on a project and what their objectives and goals are. On the other hand, we will discuss what team members may be thinking through the project stages and what their objectives and goals are.

The main tools in managing the project, which are time, resources, cost, and quality management, shall be illustrated by using practical examples from petroleum projects.

I have worked in a major rehabilitation project for offshore structures with the best in class international companies from the owner, engineering offices, contractors, and suppliers and everything has matched with the PMP guide. The project management staff and all the team members have ideal skills and competence as described in many textbooks, but unfortunately the end users are not fully satisfied, which is a situation that usually faces us in industrial projects. The solution to this problem is proposed by using a whole building commissioning system that is used successfully in an administration building and this management system is clearly illustrated in this book.

This book tries to be practical and, at the same time, match with the Project Management Professional (PMP) guide, so we selected one hundred questions from the PMP exam to help you obtain the certificate. We chose questions that present actual cases we face in managing industrial projects.

Mohamed Abdallah El-Reedy, Ph.D
Email: elreedyma@gmail.com
www.elreedyma.comli.com
Cairo, Egypt

About the Author

Mohamed A. El-Reedy's background is in structural engineering. His main area of research is the reliability of concrete and steel structures. He has provided consulting to different engineering companies and oil and gas industries in Egypt and to international companies such as the International Egyptian Oil Company (IEOC) and British Petroleum (BP). Moreover, he provides different concrete and steel structure design packages for residential buildings, warehouses, and telecommunication towers and electrical projects with WorleyParsons Egypt. He has participated in Liquefied Natural Gas (LNG) and Natural Gas
Liquid (NGL) projects with international engineering firms. Currently, Dr. El-Reedy is responsible for reliability, inspection, and maintenance strategy for onshore concrete structures and offshore steel structure platforms. He has performed these tasks for hundreds of structures in the Gulf of Suez in the Red Sea.

Dr. El-Reedy has consulted with and trained executives at many organizations including the Arabian American Oil Company (ARAMCO), BP, Apachi, Abu Dhabi Marine Operating Company (ADMA), the Abu Dhabi National Oil Company, King Saudi's Interior Ministry, Qatar Telecom, the Egyptian General Petroleum Corporation, Saudi Arabia Basic Industries Corporation (SAPIC), the Kuwait Petroleum Corporation, and Qatar Petrochemical Company (QAPCO). He has taught technical courses about repair and maintenance for reinforced concrete structures and advanced materials in the concrete industry worldwide, especially in the Middle East.

Dr. El-Reedy has written numerous publications and has presented many papers at local and international conferences sponsored by the American Society of Civil Engineers, the American Society of Mechanical Engineers, the American Concrete Institute, the American Society for Testing and Materials, and the American Petroleum Institute. He has published many research papers in international technical journals and has authored seven books about total quality management, quality management and quality assurance, economic management for engineering projects, and repair and protection of reinforced concrete structures. He received his Bachelor's degree from Cairo University in 1990, his Master's degree in 1995, and his Ph.D from Cairo University in 2000.

1

How to Manage Oil and Gas Projects

1.1 The Principal of Project Management

The subject of project management has become one of the most common themes in the recent past, and that is due to the increase in the number of mega projects worldwide and the development of modern technology in all areas of knowledge, which requires new methods of project management to cope with fast-pace developing.

Oil and gas companies are clear examples of the difference between the concept of a project and daily routine operations. These companies, in most cases, have an operations department and project department and they should work together.

Therefore, project management is different from the daily activity in operation management. Thus, most books and references that discuss project management define a project as a number of tasks and duties to be implemented during a specific period of time in order to achieve a specific objective or set of specific targets.

In operation management, production managers focus on the daily oil or gas production compared to the previous day. Oil and gas production is measured by the number of oil barrels produced per day (BOPD) in millions of standard cubic feet of gas (MMSCF). So, the monitoring of production daily is very important to the present income of the company.

On the other hand, the definition of project management can be summed up as planning, organization, recruitment, direction, and controlling of all kinds of resources in a certain period of time in order to achieve a specific objective for financial and non-financial targets.

To clarify the difference between project management and operations management, we should consider what goes on in the mind of these two managers. The project manager's goal is to finish the project on time. Then they evaluate where they will relocate after finishing the project. This is very different from the thinking of the operations manager, who never wants daily production to stop. So, they could not dream of work stopping, which is contrary to the project manager's target. Therefore, you can imagine the difference between the thinking of the two managers.

The first difference in the definition of project management is the goal to finish the project in a certain amount of time and its set of objectives all at once. While some measures are applicable for both operation and project management, the use of budget and manpower puts an end to specific actions.

1.2   Project Characteristics

One of the most important features of a project is the selection of individuals at different locations of the same company. In some international projects, the individuals are from different countries, cultures, educations, and employment and all of these individuals have different skills. With all those differences, they must work together to complete the work in a specific time and definite target.

The project manager has to coordinate between the members of the project to reach the goal of the project. As a result of rapid development in modern technology, this specialty has become important because, now days, any project contains many different disciplines. An explicit example is in construction projects, where there is a team for constructing the reinforced concrete and other teams for finishing the work, such as plumbing work. So, every branch of the construction activity has its own technology and skills. Therefore, the project manager has to cooperate between the different disciplines to achieve the project objective.

The primary goal of a project manager is to complete a project with high quality and achieve the objective at the same time.

Every project has a main driver. In general, the driver is one of the two driving forces or, in other words, is one of the two philosophies in managing a project. One of them is cost-driven and the other is time-driven.

The driver is considered to be the underlying philosophy in the management of a project, which must be determined by the director of the project with other parties, as well as the official sponsor of the project and the stakeholder. The project driving philosophy should be known to both the technical and administrative department managers.

To illustrate the above two factors' effects, we should think about all types of projects that are running around us. We will find that, in some projects, reducing the cost is the major factor and the time will be the second factor and, when the project duration time increases, it will not affect the project in the operation phase or, in precise meaning, it will not affect the owner and his investment. The building of houses, mosques, churches, museums, and other projects that have a social aspect is an example.

On the other hand, the aim of some projects is to reduce the time, which is the main challenge, so it will be a time-driven project. Examples of these projects include hotel projects because any projects that save in time will gain in profit, for an owner's profit from a hotel is calculated per day of using the rooms in the hotels. Other examples are oil and gas or petrochemical projects in the petroleum industry, where any day that can be saved will save millions of dollars per day since production is measured by barrels of oil per day (BOPD) or millions of standard cubic feet per day (MMSCFD), which will multiply the oil or gas price respectively and bring in more revenue. For example, if the gain of production from the project is 50,000 BOPD with an oil price of forty dollars per barrel, every day can save and the owner can gain 2,000,000 dollars.

From the above discussion, the main driver in petroleum projects is time. Therefore, the main target in these projects is to reduce the time. It is very important to define the basic driving force for a project, which is either cost or time. It is essential that all staff working on the project should know this information and this is the responsibility of the project manager.

Any group of teams at work, both in design or execution, should provide proposals, recommendations, and action steps that are in the same direction of the project driver in reducing the time or cost.

It is necessary that the target is clear to everyone to avoid wasting time in discussing ideas and suggestions that are not feasible. Imagine that you are working on a housing project and one of the proposals from the engineers is to use a type of cement to provide a rapid setting to reduce the time of construction, but it will increase the cost. Is this proposal acceptable? Certainly, it will not be accepted. On the other side, in the case of the construction of an oil or gas plant or new offshore platform, imagine if one of the proposals is the use of materials that are the cheapest, but it requires extra time to import from abroad, which will delay the project some days. Is this proposal acceptable? Of course this proposal is unacceptable, but if we use these proposals for the other project, we will find that the two proposals are excellent and acceptable.

It is clear that when we lose communication between the project manager and the personnel, there is a lot of confusion. If everyone works hard, but in different directions, this becomes wasted effort and everyone is not going in the same direction in order to achieve the success of the project.

Moreover, it is important to communicate with suppliers and contractors, so that their proposals in supply materials and construction should be within the project driven criteria.

Project characteristics can be summarized as follows:

- A project has a specific target.
- A project is unique and cannot be replicated with the same task and resources expecting to give the same results.
- The focus is on the owner requirements and his or her expectations from the project.
- It is not routine work, but there are some tasks that are routine.
- A project consists of a number of activities that contribute to the project as a whole.
- There is a specific time in which to finish a project.
- A project is complex in that it works by a number of individuals from different departments.
- Project managers must be flexible to cover any change that occurs during the project.
- There are uncertainty factors, such as the performance of individuals and their skills, for some of the unfamiliar work or unknown external influences that may not have happened before.
- The total cost is defined and has a limited budget.
- A project gives unique opportunities to acquire new skills.
- It gives impetus to the project manager to learn to work under changing circumstances, as the nature of the project is to change.

- There are risks with each step of the project and the project manager should manage the risks to reach the project goal at the end.

1.3   Project Life Cycle

The project definition is a set of activities that has a start time, time period, and end time. These activities vary from project to project depending on the nature of the project. For example, a cultural or social project or civil project such as the construction of a residential building, hospital, road and bridges or industrial projects are different in their characteristics. In our scope we will focus on industrial projects.

Civil projects, in general, vary from project to project depending on the size and value of the project. It can be anything from constructing a guard-room to constructing a nuclear plant.

Therefore, the quality varies depending on the size of the project, especially in developing countries.

In a small project, it might be sufficient to apply a quality control only where small contracting companies or engineering offices do not wish to have a global competition. For example, increasing the quality will increase the project's total cost and if these companies have quality assurance tracking systems that will also increase the cost of the project as a whole. Therefore, they often apply the quality control only within the structure safety of the building.

In the case of major projects, there are many execution companies or engineering offices working. Therefore, we must also take into account that firms implementing quality assurance procedures are necessary and vital, as well as the quality control carried out in all phases of the project based on the project specifications.

Stages of construction projects start with a feasibility study, followed by preliminary studies of the project, following detailed studies with detailed drawings. Then, the operation crew will receive the project to run.

In all these stages, there are many types of quality control that are required to obtain a successful project that can return benefits and money to the owner and all participants in the project. Figure 1.1 shows the life cycle of all projects.

From this figure, it is clear that when a feasibility study has been finished five percent of the progress of the project is shown and, upon completion of the engineering designs, 25 percent of the project progress is shown,

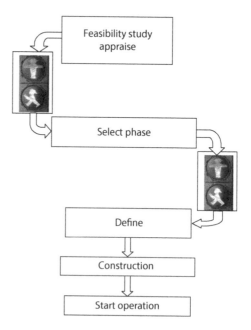

Figure 1.1 Project life cycle.

and the biggest project stage, in terms of time and cost, is the period of the execution phase.

As shown in Figure 1.1, after the feasibility study, senior management should have a definite answer for the following question: will the project continue or will it be terminated? If there is a positive situation, then cross the gate to the next stage which is preliminary studies, which will provide a more accurate assessment of the project. After that, another decision will need to be made on whether the project will move forward to the detailed engineering and construction phase.

At each phase of the project, there is a role for the owner, the contractor, and the consulting engineer. Each system has its own method of project management and every stage of these methods has its own characteristics and circumstances. These follow a change in the area of employment or Scope Of Work (SOW) that clarifies each stage for each of the three parties.

A characteristic of the project life cycle is that it changes from time to time. In each period there is a different number of personnel and employment in the project. For example, at the beginning of the project, the number may be very small but then increase when increasing the number of activities carried out and then gradually decrease until the end of the

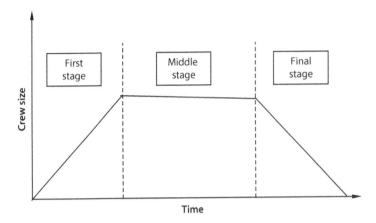

Figure 1.2 Change of crew size during project life time.

project. Figure 1.2 shows the change in the number of personnel in the project.

From the above figure, it is noted that the project manager should have the necessary skills to deal with the changes that occur during the life cycle of the project.

1.3.1   Initiation of the Project

In any major projects there is an involvement of many project managers as there is an owner, an engineering consultant, and a contractor. They should all go through the same steps that we will discuss, but each person does it based on his or her goals, target, and company system.

In general, any project starts from a creation of a formal document called the project charter. The project charter is described in the PMP guide, but its name is different from one company to another. This document is extremely important for getting a project started in the right direction.

There are many reasons for starting a project. In general, for commercial and industrial companies, making money is the reason for doing a project. However, in some cases, there are many other reasons for doing projects, such as to follow government regulations and laws, to enhance the health, safety, and environment (HSE) for a company, or, more specifically, to help with oil disposal and the instant cleaning of the Gulf of Mexico due to the oil spill that happened in 2010. In some industrial and commercial companies, the projects stay current with developing technology.

A project charter is defined in PMPBOK and is expanded in the third edition due to the importance of this paper. It also recommends that the contract with the customer will be completed before the approval of the project charter.

Noting that, the definition of the customer has a wide range as everyone, including the project managers, are a supplier and customer at the same time.

When the contract is signed by the customer, the scope of work and deliverables should be clear, as the change will be very limited after the contract is signed. Therefore, there will be enough information to be included in the project charter.

The definition of the project charter in PMPBOK is a document that formally authorizes a project and includes directly, or by reference, other documents as the business needs the product descriptions.

This document is usually made by the senior project manager, as the project manager will not be defined in this stage, so the document should be simple, precise, and accurate. Putting the reference is not recommended because top senior management does not have time to go into depth in the document. Also, I agree with Newell (2005) that this document should be small. If it is a big document you will face many inquiries.

This document is usually contains the following:

- The name of the project
- The purpose of the project
- The business need for the project
- The rough time schedule as defined by the project time period
- The budget of the project
- The profit from the project using the payout method (discussed further in Chapter 3)
- The project manager in any situation

After signing this document, the project manager will be selected through a discussion between the project sponsor and the senior managers. In the case of a small project, the project manager has been defined, so there is no need to include his name. In addition, the project manager will prepare this document under the supervision of the project sponsor.

It is better that the project manager prepare this document, as he or she will be the most involved in the project and will closely understand the target and goals for the senior manager.

1.3.1.1   Getting to the Scope Baseline

As previously discussed, everyone in the project is a customer and a supplier at the same time, including the owner who is a supplier to the operation department in his company or any other user.

The key topic in any contract between two parties is to define the scope. As defined by PMPBOK, the term scope may refer to the following:

- Product scope, which includes the features and functions that characterize a product or service
- Project scope, which is the work that must be done to deliver a product with the specified features and function to the end user

The product, which will be delivered through the project, should satisfy both the customer and the stakeholder.

The scope should be prepared after clearly defining all the stakeholders. Take more time in this stage, as at the end you should define the scope baseline in any way that will not finish in days, but in weeks and months. Feel free to take any idea into consideration, especially from the key persons sharing in the project. The last thing you need is, after finishing the scope of work, someone saying "we need to add some scope" or "we need a little change".

So, after many meetings, reduce the unnecessary items from the scope or define part of the scope to the supplier so that the scope baseline is documented and approved by the concerned stakeholder.

After you define the scope of work, be sure it is clear to the supplier who will provide this service. You should use any communication and skills necessary to make the scope of work clear to the supplier. An engineering company will provide a list of deliverables. After you send the company the scope, be sure that the deliverables match with your requirements and that everyone has read any statement based on his or her background and previous experience.

It is better to return to similar projects and look at the work break down structure (WBS) and then review if you are missing anything from the deliverables list. In major projects, every discipline should review the deliverables list received.

This document needs to be clear because many people will read it. The SOW is the major main part on the statement of requirement document (SOR) as most of the conflict in any project is due to misunderstanding the scope of work. In some cases, the supplier may provide a small user manual to use for maintenance service. On the other hand, the operation and maintenance engineers may be waiting to receive a comprehensive user guide as they have a full responsibility to do the maintenance in house and avoid using the supplier in minor maintenance situations based on their policy. In some cases, they are afraid that the supplier will be out of business or has merged with another company, which traditionally happens. After receiving this manual, you may be in crisis because the supplier is doing what you are requesting, but the end user is not satisfied. In this case, you will change order. From this example in the deliverable list, the contractor will deliver a "user manual", but it is different from the perspective to the stakeholder expectation.

This situation is repeated many times in oil and gas projects. However, if we apply the whole building commissioning system methodology, as presented and discussed deeply in Chapter 8, these problems may not occur.

The acceptance criteria, the test procedure, and criteria should be defined in the scope of work. So try all of the deliverables that are tangible and measurable items that can be easily understood.

1.3.2   Feasibility Study

Each phase of a project has a different importance and impact on the project as a whole, but each phase differs depending on the nature, the circumstances of the project, and its value and target.

The phase of the feasibility study is the second step after the emergence of the idea of the owner. The owners in an oil and gas project are the geologist and petroleum engineering team, whose idea is based on oil and gas reservoir characteristics.

The economic study for the project will be performed by personnel from a high level of the organization and with high skill, as this study will include the expected fluctuation of the price for oil and gas and other petrochemical products during the project lifetime. Their experience is based on similar, previous projects so they have records and lessons learned from the previous projects.

In this initial phase, the selection of the team or the consultant office that will perform this feasibility study is important. In some cases, there may be input from an engineering firm to perform a generic engineering study about the project and estimate the cost based on their experience.

This is calculated by using quantitative risk assessment by using the Monte-Carlo simulation technique, as it needs to take a decision for a big decision.

There are many factors to consider in this study. For example:

- The agreement between the partners
- The expectation of the oil and gas price trend
- Political situation for the country that has the products
- Government laws, regulations, and taxes

The phase of the feasibility study is matched by the technical project stage, which is appraised for the preliminary (FEED) study phase. These two phases are very essential because they set the objective of the project and identify engineering ideas through the initial studies. It is preferred to apply the Japanese proverb, "think slowly and execute quickly," especially in the feasibility study stage. This is the stage of defining the goal of the project and determining economic feasibility from this phase should determine the direction to move also.

For those reasons, we must make this phase take advantage of its time, effort, study, research, and discussions with more care about the economic data.

The economic aspect is important at this stage, but the engineering input is very limited.

1.3.3   FEED (Preliminary) Engineering

This stage is the second phase after the completion of the feasibility study for a project.

This phase of preliminary engineering studies, which is known as feed engineering, is not less important than the first phase.

This phase of engineering is one of the most important and most dangerous stages of engineering and the professionalism of the project since the success of the project as a whole depends on the engineering study in this phase.

Therefore, as this stage is vital, the engineering consultancy firm that performs this study should have strong experience in these types of projects.

For example, a Liquefied Natural Gas (LNG) project is a type of project that needs an experienced office. Another example would be offshore projects that use FPSO and that also need a special consulting office that has worked on this type of project before.

In the case of small projects, such as residential or administrative buildings or a small factory, the phase of feed engineering is to deliver the type of structure, whether it would be a steel or concrete structure. If a concrete structure is decided, the engineer should define it as a pre-cast concrete, pre-stress concrete, or normal concrete and then determine the type of slab structure system as a solid slab, flat slab, hollow blocks, or others.

Also, this phase defines the location of the columns and the structure system and if it will use a frame or shear wall for a high-rise building.

In summary, the preliminary engineering is to provide a comparison between these alternatives and the variation depending on the size of the building itself and the requirements of the owner. The reasonable structure system and similar mechanical or electrical system will be selected, so this stage is called a select phase.

In the case of major projects, such as a petrochemical plants or new platforms, there will be other studies in this stage such as geotechnical studies, met ocean studies, seismic studies, and environmental studies.

The main element of this study is to provide the layout depending on the road design, location of the building, and hazard area classification in the petroleum projects.

Moreover, it needs to select the foundation type, if it is a shallow foundation or driven or rotary piles based on the geotechnical studies.

In case of oil and gas projects, we need to carefully study the mode of transfers and trade-offs of the product and select the appropriate methods of transferring between the available alternatives options.

Now, it is clear that, as a result of the seriousness of that stage and the need for extensive experience, in the case of large projects, the owner should have competent engineers and an administrative organization that have the ability to follow up on initial studies in order to achieve the goal of the project and coordination between the various project disciplines, such as civil, mechanical, electrical, and chemical, as all the disciplines usually intersect at this stage.

Generally, regardless of the size of the project, the owner must prepare the Statement Of Requirement (SOR) document during the preparation of engineering requirements. The SOR will be a complete document containing all the owner information and needs to be about the project with high accuracy and contain the objective of the project and the needs of the owner precisely and identify the scope of work (SOW).

This document is a start-up phase of the mission documents quality assurance system, as this document must contain all that is requested by the owner. The SOR document must outline the whole project and have a document containing all particulars of the project and its objectives and proposals and the required specifications of the owner.

Title:-	Statement of Requirements (SOR) Preparation
What	The SOR is a formal document. It can vary from being a one-page document (minor projects) to a sizeable document incorporating the "basis of design," i.e., plant, pipe sizes, pressures, etc.
Why	The SOR is intended to document, in a clear and unambiguous manner, the key engineering inputs and the major engineering requirements and management tasks that have to be completed in order to meet a particular business objective, this objective being clearly defined at the beginning of the SOR. The completed SOR is intended to identify the factors that the business sponsoring the project considers important to the ultimate success of the project, as well as being a high level specification of project deliverables.
How	Create a formal document, depending on the project needs.
When	Within the project framework, the SOR will form is an integral part of the select stage DSP and it is required for the chosen option at the end of Selection. The project should not continue into Define until the SOR has been approved.
Who	In practice, the SOR is usually prepared by the project personnel who liaise closely with the business unit personnel (SPA). It is important that the BU formally approves the SOR as it is effectively a contract between the BU and the project team defining high-level deliverables and expectations. Similarly, because of the significance, a change management procedure should be established that will ensure all the changes receive the necessary approval.

This document also contains the technical information available from the owner, such as the location of the land and its coordinates system and its specifications. Noting that, this document will be a part of the contract document between the owner and the engineering firm, noting that the engineering firm will provide the cost time and resources (CTR) sheets based on this document.

In the case of projects such as gas, an LNG gas liquefaction project determines the amount of gas, type, and specifications which are needed to process and transfer with the clarification of temperature, pressure, and all other technical data for the final product to be shipped or transported outside. This is one of the most important data to be mentioned in the document, as it identifies the project lifetime.

Specifications required by the owner in the project should be defined clearly and precisely in this document.

It should be noted that we must hold many of the regular meetings between the owner, technical team, and the consulting engineering responsible for the preparation of initial studies. Through that, the SOR may be amended several times and each time the document must contain the date and revision number to contain all of the requirements of civil, architectural, electrical, mechanical, and others found in the project.

We may recall here that in quality assurance we must be sure that the final document resides with all the parties and that everyone is working through this document. This must be done to determine the number of meetings and the exact schedule of meetings needed to reach the target required.

The SOR document is not only required for the new project, but it is also needed in the case of modification to the buildings or in the plant. In the case of small buildings, the owner should define the required number of apartments, floors, and stories or any other requirements the owner feels are a benefit to his target.

Upon receipt of the engineering office, the SOR document is to respond to the owner document with another document that is called the basis of design (BOD). Through the document, the engineering firm will clarify the code and engineering specifications, which will operate in the design as well as the calculation methods, theory, and computer software that will be used.

This document may state the required number of copies of the drawings that will be sent to the owner and the sizes of those drawings.

In addition, the engineering firm should request any missing data and require a third party to supplement information such as weather and environmental factors. This document will be reviewed by the owner carefully and can be amended until it satisfies the two parties.

At this stage, it is important to make sure that both the owner and the engineering firm have the same concept and there is a complete agreement among all the technical aspects. In the preparation of any drawings, we are now in the FEED studies, in which the drawings should be delivered to the owner to review and give input. The owner and the engineering firm should agree on the number of reviews of the document, and, if it goes over the specific time, it is assumed that the owner has accepted it. This is very important in controlling project time.

This phase may take a number of months in the case of large projects, and, therefore, the technical office of the owner must have a qualified engineer with experience for controlling costs and follow-up time according to the schedule agreed upon in advance. We may need to consult a specialized engineer in planning who is the Planner Engineer.

The engineer should be specialized in cost control, the estimated cost of the project, and the expected time, which is comparable in the feasibility study.

With the passage of time, clearly selected the equipment, and the project layout in its final stage, the project cost estimate will be more precise until things are nearer to the end of the initial studies. Then, one can obtain the nearest possible accuracy of the cost of the project as a whole.

It is worth mentioning that in investment projects, such as petroleum projects, savings in time bring a big return for where the return of income or expense is calculated by the day.

It's imperative that we note here that, at this stage, one should not overlook the way in which to determine the maintenance of the buildings' and the facilities' foundation in oil and gas plants in the future, which can be done by establishing the age of the structure and defining the structure lifetime, type of structure, and the ways of maintenance. The project site itself and the surrounding environment must be considered to determine the ways to protect it from weather, reducing the cost of maintenance over time by selecting different methods of maintenance.

For example, you can protect a reinforced concrete foundation from corrosion by protecting the reinforcing steel, for example, through a system of expensive protection at the beginning of the construction with periodic low-cost maintenance.

On the other hand, we can use a low-cost alternative during construction and high-cost regular maintenance as a simple example if we don't use any external protection system.

The structure, the mode of operation, and the maintenance plan all have an impact on the preliminary design.

For example, in power stations we must ask whether the water tank can be repaired, maintained, or cleaned. To answer this question, you must decide if it needs additional tanks as standby for maintenance purposes or not.

In this phase, many other initial design decisions must be made, and therefore this stage, as previously mentioned, requires high experience, since any error would lead to major problems during operation, which could cost a lot of money and could be prevented by a low-cost solution in this phase.

1.3.4   Detail Engineering

At the end of this phase, the engineering office will deliver the full construction drawings and specifications for the whole project that contain all the details that the contractor will execute.

In this phase, there will be a large number of engineering hours, so it is necessary to have good coordination between the different disciplines.

Changes might occur in the cooperation between departments and, therefore, vary depending on the performance of the work of managers. The system of quality assurance is a benefit as it provides us with the basic functioning of all departments, despite any change in personnel. These problems often occur at the stage of the study that requires the cooperation of extensive, vital, and influential team members among the various departments of engineering such as civil, architectural, mechanical, and electrical departments.

For example, when the managers of the Department of Civil and Mechanical Engineering have a strong relationship and there are regular meetings, good work will come from it and the meetings and correspondence will be fruitful.

You can easily determine whether your business might benefit from the quality assurance systems or not by taking a closer look at past experiences (and this method has been described in the book by Richard German).

When you discuss with the team and you find that the goals and expectations are not clear between colleagues, then you should ask yourself if the majority of problems can be avoided if individuals committed to work through pre-agreed measures. All colleagues should fully and clearly know their role.

The system of quality assurance in this stage is important as it organizes the work, everyone knows the target of the project, and everyone's responsibility in the project is clear. The concept of quality is defined with a supported document.

The documents are regarded as the executive arm of the process of the application of quality. Therefore, any amendment or correction in the drawings should go through the procedure and agreed system.

The drawings should be sent in a specified time to the client for review and discussion through an official transmittal letter to control the process time. Any comments or inquiries should be discussed and made through agreement between the two technical parties, then the modification will be made by the engineering firm and resent to the client through the same communication procedure.

Through the development of a particular system, to avoid other copies of the drawings, so as not to confuse with them and to avoid human error, through control, the documents register and should have the date of the drawings, engineering review dates, and continued monitoring in the document cycle until it reaches the final stage of the project. Finally, the final approval of the drawing will be taken and sealed with stamp indicating "Approve For Construction" that these are final drawings for approval of the construction.

After the completion of the detailed engineering phase, the specifications and drawings are ready for the execution phase. You can imagine that, in some projects, the documents may reach hundreds of volumes, especially the specifications and other operation manuals, as well as volumes of maintenance and repairs.

1.3.5 Decision Support Package

This package of documents is usually put together by the project manager and the project team, then it's presented to senior management to help them make decisions, as shown in Figure 1.1. A decision will be made at every project gate in order to enter another stage and it is very important after the feasibility study and feed engineering phase.

In order to define the exact DSP and how to implement it, the following questions should be answered:

What is the DSP? The Select Decision Support Package (DSP) is a compilation of key project information used to support decision-making at this gate. The decision to be made is generally whether or not to fund the Define stage of the project. Therefore, the DSP must accurately support the team's recommendation with particular emphasis on potential rewards and risks. The CVP stipulates that a project should not progress through the gate to the next stage until the project team presents the Gatekeeper with the "key to the gate" -- the Decision Support Package (DSP). This document should include information taken from the Select stage activities that are necessary for the Gatekeeper to review and approve the project for the next stage. A plan for the next stage incorporates the following:

- A clear set of expectations
- Signed SOR
- PEP (including project specific WBS)
- Conceptual factored Class 3 estimate
- Holistic risk assessment
- Exit strategy
- Output from applicable VIP
- Defined list of capital and manpower resources for the next stage

This decision should be aligned with the business goal, strategy, and objectives and it must be determined based on the needs or deliverables of the appraised CVP stage.

The DSP includes three major components: the Executive Summary, the DSP Notification Document, and the DSP Reference Document.

The Executive Summary is a stand-alone document that provides an overview of the project. The Executive Summary may range from just one page to ten or more pages, depending on the size of the project. The Executive Summary includes:

- Project Overview (includes SOR)
- Business Case
- Decision & Risk Analysis
- Plan for Project and Next Steps

The DSP Notification Document provides a short formal written (or electronic) record summarizing the opportunities or options to move forward into the next stage, together with documentation addressing those opportunities or options that are being dropped. It is intended to be shared with all stakeholders of the project.

The Reference Document for DSP contains all other reference materials such as project schedule details, contract work scopes, etc. which have been succinctly presented in the Executive Summary. This documentation is retained as reference material, and it isn't formally distributed outside of the project team.

Why is DSP important? The Select DSP allows the Gatekeeper to make an informed decision as to the next course of action in relation to any specific project, i.e., the key to the Define gate. It will provide information on the best-identified project approaches and analyses concepts, and it will include prelim cost estimates to confirm project viability in line with the business strategy. In addition, one of the most important uses of the Decision Support Package is to ensure that the right people are selected for the next stage of the project, even before you get to that stage.

Using information provided in the Select DSP, the Gatekeeper either:

- approves the project, giving the team the ability to pass through the gate to the next Select stage;
- defers the project, based on portfolio management;
- "kills" the project; or
- recycles the project.

Keep in mind that "recycling" a project back through a stage should be a rare occurrence, and it is not really a desired option. A recycled project often indicates a failure in communication between the Gatekeeper and the project team.

The Select Decision Support Package (DSP) is a compilation of key project information used to support decision-making at this gate. This is a formal document that will be issued and presented to the Gatekeeper for review at the end of the Select Stage.

The project should not progress through the gate to the next stage until the project team presents it to the Gatekeeper, who is usually the senior manager with the "key to the gate," which is the Decision Support Package (DSP). This applies at all gate stages within the project life cycle.

Single Point Accountability (SPA) for the project, who is usually the overall project team leader, should deliver the Select DSP in line with Gatekeeper's expectations. They should be assisted where required by appropriate resources needed to provide overall project assurance as well as an increased involvement by the project teams. It is essential for the project's success that the correct team is formed to deliver the select deliverables and DSP. The select stage of the project should not proceed unless there is a clear business commitment to these deliverables through resource allocation and support. The project lifecycle commences at the beginning of selection.

1.3.6   Design Management

The target is to control the design stage to provide high quality with a better price. The design input is all the technical information necessary for the design process. To be clear, the basis of this information comes from the owner through the statement of requirement (SOR) document, so the engineering firm should review this document clearly and, if there is any confusion or misunderstanding, it should be finished and clarified in the document and through meetings.

Instructions for controlling the design are often provided in the contract. The client puts in some instructions to control the whole process or request some specific action, such as a representative from the audit during the design phase.

The designer must take into account the available materials in the local market of the project country and its location and they must match with the capabilities of the owner. The designers must have contact with and full knowledge of the best equipment, machinery, and available materials.

The design must be in conformity with the project specifications and the permissible deviation and tolerance should be in accordance with the specifications and requirements of the owner.

Health, safety, and the environment are critical subjects nowadays, so every design should match with health, safety, and environmental regulations.

The computer is one of the basic tools now in the design process as well as in the recording and storage of information, with the possibility of changing the design easily. It is easy now to modify the drawings by using Computer Aided Design (CAD) software in order to obtain more precise information with the access to information through various forms of tables and diagrams.

The design output must be compatible with all design requirements, and the design should be reviewed through the internal audit. The design must be compared to an old design that has been approved for similar projects. Any engineering firm should have a procedure, such as a checklist, to review the design.

The audits of design review are intended to be on a regular basis. In the case of important stages in the design, the audit must have complete documentation and could take the form of analytical forms such as the analysis of collapse, with an assessment of the risk of Failure Mode and Effects Analysis (FMEA). In the case of oil and gas projects, operating risks such as Hazard Operation (HAZOP) are being studied. The review will be conducted by engineers with higher experience.

1.3.7 Execution Phase

Now everything is ready for this stage, and this stage involves both quality assurance and quality control, especially in the reinforced concrete works and in the concrete itself composed of many materials such as cement, sand and coarse aggregate, water, and additives in addition to steel bars. Therefore, it is essential to control the quality of each element separately, as well as the whole mixture. In addition to that, the quality control should follow restrictions during the preparation of wooden form, the preparation and installation of steel bars, and pouring and curing concrete.

It is clear here that the contractor should have a strong, capable organization, capable of good quality control, and documents that define the time and date in which the work was carried out and who received the materials. At the same time, it's important to determine the number of samples of concrete and define the exact time, date, and result of each test.

Often, during execution some changes occur in the construction drawings of the project as a result of the presence of some of the problems at the site during the construction or the presence of some ideas and suggestions that can reduce the time of the project.

However, it is important that the change of work be done through documents in order to manage the changes and build them into the drawings.

The supervisors and the owner must have their special organization. The owner organization in most cases has two scenarios:

- The owner will establish an internal team from the organization to manage the project; or
- The owner chooses a consultant office to manage the supervision on site.

In most cases the design office will do the supervision.

The construction phase shows the contractors the capabilities for local and international competition. If, and only if, the concept of quality assurance of the contractor project team is very clear and has experience in a comprehensive quality system because the aim of all the competitors on the international scene, since the period is working through an integrated system, is to confirm the quality of the work and quality control in all stages of execution in order to achieve full customer satisfaction.

1.3.8   Commissioning and Startup

Importance of this stage varies depending on the nature and size of the project itself. In the case of housing projects, commissioning and start up will be applied to the building, for example during the finishing stages as per each floor or by each stage of completion as received for the HVAC, plumbing system, painting, and other final stages to start using the building.

A team formed by the owner and consultant is to receive the work in this stage. The work will be a list of the parts that need repair from the contractor.

It is a different story in the case of industrial projects such as constructing pipelines, pumps, and turbine engines, or a new plant. In this case, a new team will formulate consisting of members of the projects and operating personnel who receive work and have the head of the team. Noting that, this team should be competent and have previous experience in commissioning and startup. The team will have a specific target to start the operation where the reception is not performed until after the primary operation. Start up and commissioning at this stage is to make sure that all the mechanical systems work efficiently and safely without any leakage or error in the operation. This stage takes a period of time depending on the size of the project and may extend to months.

The cooperation between the operation and project team is essential. The operation starts according to the schedule for this specific stage. It can be in hour units and all the parties should agree on this schedule to provide a smooth transition in a safe manner because increasing temperatures and pressure without previous study may cause a disaster. So again, competent people and a good schedule are the keys for successful commissioning and startup.

1.4   Is this Project Successful?

When determining whether a project successful, you should focus on the management of the project. You might see a high-rise building and think that the project seems successful, but is the project management successful too? To answer this question you need to answer the following three questions:

1. What is the plan and actual execution time?
2. What is the actual cost and budget?
3. Is the project performance according to the required specifications?

For the last question, we can answer yes because for a big project we cannot agree or approve anything with lesser quality or that is beyond specification. It is safe to assume that the quality is a red line that cannot be crossed or negotiated.

Therefore, the successful project manager has to achieve the goal of the project and satisfy all stakeholders. At the same time, the completion of the project on time and cost will not be more than the approved budget.

1.4.1   Project Management Goals

Every project should have a specific target or targets and each target should be defined for all the team members in the project.

In any project, we will have the following elements:

- Money
- Manpower
- Machines

The target of project management is to use the above elements to achieve the project with less cost and high quality.

So, every job in the project must manage one or more of the available resources and optimize the use of the resources in order to reduce losses and achieve the project target within the time constraint, cost, and quality.

Based on the PMP guide, project management is characterized by multiple areas or topics that must be managed at the same time. These topics are as follows:

1. Project integration management
2. Project scope management
3. Project time management
4. Project cost management
5. Quality management
6. Human resource management
7. Communication management
8. Procurement management
9. Risk management

Based on Kotter, J.P (1996), there is a good differentiation between the management and the leader. Management is a set of processes that can keep a complicated system of people and technology running smoothly. The most important aspects of management include planning, budgeting, organizing, staffing, controlling, and problem solving. Leadership is a set of processes that creates organization or adapts to significantly changing circumstances. Leadership defines what the future should look like, aligns people with that vision, and inspires them to make it happen despite the obstacles.

1.4.1.1   Project Integration Management

The purpose of integrated management is to ensure that all elements of the project that are interdependent have a good correlation between them. It is done through good planning and the existence of an operational plan for the project, which is the interdependence between the members of the team. At the same time, it finds a way to control the project performance in case of any change in the project.

1.4.1.2   Project Scope Management

The extent of the project, its size, and what must be fixed and known need to be established. Therefore, we must take the necessary actions to ensure that all work required has been known with specific planning and that only the work necessary to achieve the success of the project is done. This is done by identifying the volume of work and planning.

1.4.1.3 Project Time Management

The project must have a specific start and finish date and the duration of the implementation of the project must also be defined either by the owner or by the contractor who will carry out the activity or by the agreement between both.

So that we can finish the project according to the schedule and manage the project time, the schedule plan must be done for all the activities and the time of each activity and the relation between them must be defined.

No one can achieve the project on schedule if the time commitment for each activity is not followed.

1.4.1.4 Project Cost Management

To determine the time required at each of the activities of the project, the resources should be allocated to the execution of the activity by the individuals, materials, and equipment.

Therefore, the cost of each activity and the rate of spending on each activity have to be estimated to obtain the cash flows through the project.

Managing the cost of the project requires actions and steps to ensure that, in the end, the total money spent on the execution of the activities equals the budget allocated for the project, which will be allocated from the initial cost estimate.

1.4.1.5 Project Quality Management

There is no doubt that the quality borders are a commitment to quality specifications, both for the materials and processes. In most cases, quality is overlooked when discussing the projects and only focusing on time and costs.

However, it does not mean that quality is less important than time and costs, but the relationship between these three elements are very dependent on each other. So, to be clear, any activity is considered to have been finished only if it matches the specifications. There is no doubt that these specifications have a significant impact on determining the cost of the activity.

It is clear, already, that the management of a project is done through a set of plans for the time, cost, and quality. The preparation and implementation of these plans should be shared with all the team. It is normal that any failure of any function of the organization affects the entire project, in terms of time or cost or quality.

1.4.1.6   Project Human Resource Management

It is essential to use ordered steps and reasonable tools that allow making good use of manpower. The establishment of a good organization, according to the needs of the project and the needs of individuals, is essential. The work of human resources should match with the schedule planning of the project. There must be conformity between the individual objectives and project objectives through motivation and knowing that the continued success of the project is a success for all. To be profitable and gain a strategic win is the basis for the interaction between project management and personnel.

1.4.1.7   Project Communications Management

There must be a procedure for providing good communication between members of the team by planning and holding regular meetings for team members and other partners in the project.

The way of transferring information between the team members should be identified, quick, and easy. Now, there are many types of technology that can achieve management and good communication such as networks, e-mail, and meetings that can be held through a video conference.

1.4.1.8   Project Risk Management

Each project has its own risk, whether the result of technical aspects or the result of procedures and the project execution sequence. Therefore, the risk should be managed by identifying it first, setting priorities, and then finding a solution according to the type of event, its likelihood of occurring, and its potential impact on the project.

This is followed by a periodic follow-up phase precisely with the assurance of the distribution of responsibilities and authorities for each item that will have a high impact on the project.

1.4.1.9   Project Procurement Management

Every project depends on the procurement of materials or equipment. Therefore, it is necessary to prepare procedures to deal with external vendors to make the purchases serve specific objectives of the project.

Therefore, we must determine purchasing strategies as well as the nature of the contracts and how to manage the identification and follow-up of procurement procedures. We must also identify how to maintain the quality of purchases or services provided to the project.

1.5   Project Management Tasks

Every manager in the project organization is responsible for planning and monitoring the plan and assuring that the executive work matches with the plan. To achieve that, he should coordinate with other managers and provide a report to the project manager.

The main items in the planning and monitoring process are the following:

- Define project objective
- Define the work
- Define the work time period
- Define the available and required resources
- Define the cost
- Review and evaluate the master plan
- Accept the master plan
- Follow up execution
- Follow up cost
- Compare between actual work, cost, and master plan
- Evaluate performance
- Predict and change strategy

The definition of project management is illustrated in Figure 1.3. The first thing in management is to identify the target of the project management process as in the planning, execution, and follow-up. There are two factors that affect and/or are affected by resources: time and funding.

The following is a summary for all the tools available to the project manager for managing the project for success.

1.5.1   Define the Project Target

Identify the objectives of the project in the first phase of the planning stages. The project objectives must be defined from the beginning, such as the completion of the project in less time or a reduced cost.

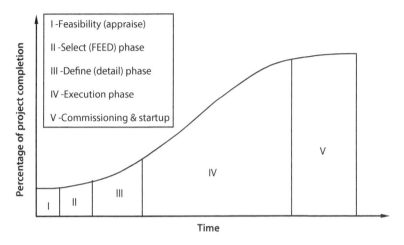

Figure 1.3 Industrial project main phases.

For projects on the technical side, we must identify the possible output or final words of the basic components of the project.

1.5.2 Define the Scope of Work

The required work should be defined precisely as it is in the basic plan for the project in three main areas: determining the work required, the resources and budgets for the work, and the time required for the implementation of the work. So, it is clear that the success of the planning process as a whole depends, to a large extent, on the identification of the work.

1.5.3 Define the Time Frame

One important factor in determining the objectives of the project is to develop a plan for the project through a time schedule. It should be developed in the form of a base scale of the project, and the main objective of the base scale of the project is to put the project time in clear steps by preparing the calendar, which is one of the main pillars in the management of any project.

1.5.4 Define the Available Resources

The basic tool for identifying resources required is a list of work that outlines the required labor, materials, and other services for each of the activities and our numbers if we have a schedule for these actions. Then, you can have a clear idea of the necessary resources at different periods. At the same time, it is required to compile information on available resources and plan work required to achieve the project.

1.5.5   Define the Cost

The cost can be determined by the following:

- The quantity of resources required to implement any action
- Cost rates for each supplier of the resources
- The time of each work activity
- Fixed costs based on the activities

Therefore, the cost of the project varies according to the size of the project and, consequently, the time of the implementation of any work depends on the time scheduled for the project.

Therefore, the budget for the project should agree with the rate of spending on the project in relation to the schedule prepared by the project team.

1.5.6   Evaluate the Master Plan

Upon completion of all the previous steps, we evaluate the master plan of the project and how it will achieve agreed objectives of the project in terms of time and costs.

There is often a need to amend the plan, often in the field of resources, as they usually clear the requirement for a certain type of a large amount of resources in one specific application. Accordingly, it will be in the line of the settlement of the rescheduling of resources of some activities, which does not affect the master plan of the project. At the beginning, the schedule must be made through discussions between the owner and the contractor. Then, the timetable will be agreed upon by all parties to achieve the success of each party according to its objectives.

1.5.7   Accept the Master Plan

Store the master plan securely as it will be the basic reference in the future. After the approval, storing it is an important step for the planner, who is to keep the original master plan without any change to return to it if there is any confusion.

Review the execution plan and the performance through this plan, as there usually will be changes in this plan.

1.5.8   Schedule Follow Up

After all the previous steps, the execution phase will start. Then, start tracking the progress of work by registering the amount of work done and the used resources. This will apply to the activities that took place as well as the activities under the operation.

Follow-up schedules are adjusted periodically in order to be commensurate with the actual on site. We must, therefore, do the follow-up periodically and as agreed upon at the beginning of the project.

1.5.9   Cost Follow Up

You should follow-up on the cost periodically during an agreed period of time to track the project cost and compare it to the estimated cost of the project that was identified through the budget and approved by the owner.

We must follow the paid cost and the cost that is due to others with the time of the purchase orders and contracts of employment. Any deviation gives a snapshot of the position to evaluate the deviations at the same time, and then we can expect the total cost at the end of the project.

1.5.10   Comparing Between Actual Work and Master Plan Cost

You must follow-up on the progress of the work. Follow-up costs in themselves do not represent the control of the project, and the control of the project includes several steps that will lead to taking the steps geared towards the achievement of the objectives of the project.

The first step is to compare the progress of the work plan. It is clear that the most important indicator is the date for completion of the project calculated by the critical path. If there was any delay in activities on the critical path, it would inevitably lead to delays in the date of the completion of the project.

You must also monitor the activities rather than the critical path, as any significant delay with low performance rates may lead to influence on the critical path.

For the activities which finish, it would be sufficient to compare the actual costs to the estimated budget for each activity.

1.5.11   Performance Evaluation

There are two main important indicators used to assess the performance of the project. The first is the date for the completion of the project and the second is the cost of the project and is not intended, here, as the use of these indicators on the activities of a separate or specified period of time. It is only intended to follow the general direction of the project through these indicators on relatively long periods of time. It is not a dangerous imbalance in one or some of the activities, but the big problem is a deviation in the general direction of these indicators.

1.6 Project Manager Skill

From the previous discussions, it is obvious that the main player in any project is the project manager, as he carries the load of formulating the team.

When thinking about selections of the project manager, consider that this role is different from operations and routine work as the project is unique and may not have been done before.

The project manager must be flexible, because the project changes from time to time. For example, at the beginning of the project there is a small number of individuals. Then, the number increases with time and, when the end of the project is near, there will be less than before.

Therefore, the behavior of the project manager must be flexible according to the variables of the project.

The project manager deals with all levels and different individuals from various departments and other parties. The project manager must be a good listener and must be able to guide and persuade.

He or she must have previous experience in the same type of project and have the necessary technical information to manage the project in an efficient manner. In addition to that, he or she must have the skills of project management in terms of managing time, costs, resources, communications, and contracts.

The skill of communication is very important, as the facts do not always present themselves. The best ideas in the world would not be known without communication. Therefore, communication and providing specific and clear instructions is important, as any misunderstanding can cause a loss of time and money at the same time. On the other hand, communication with the higher levels is necessary in order to provide summaries of the performance of the project in a way that allows senior management to help with the project and not vice versa. Finally, the project manager must have the ability to have a more accurate sense of a broad vision for the whole project. He or she must have the capability to manage dialogue, especially in meetings, which is regarded as the strongest means of communication to reach the goal that is the main success of the project. The following are some inquiries about a case study that happened in real life and these questions are the same in the PMP exam. For the answers go to the website www.elreedyma.comli.com.

Quiz

1. You are a project manager on an international project of great importance to the client. The client is from another country and is so excited by how well the project is going that he presents you with a company automobile for your personal use. The BEST thing for you to do would be to:
   - thank him and offer a gift in exchange.
   - politely turn down the gift.
   - ask that the gift be changed to something that can be shared by the team.
   - ask for a gift that can be used before you return home.

2. All of the following are part of the team's stakeholder management effort EXCEPT:
   - Giving stakeholders extras
   - Identifying stakeholders
   - Determining stakeholders' needs
   - Managing stakeholders' expectations

3. Your management has decided that all orders will be treated as "projects" and that project managers will be used to update orders daily, resolve issues, and ensure that the customer formally accepts the product within 30 days of completion. The revenue from the individual orders can vary from US $100 to US $150,000. The project manager will not be required to perform planning or provide documentation other than daily status. How would you define this situation?
   - Because each individual order is a "temporary endeavor," each order is a project - this is truly project management.
   - This is program management since there are multiple projects involved.
   - This is a recurring process.
   - Orders incurring revenue over US $100,000 would be considered projects and would involve project management.

4. Who determines the role of each stakeholder?
   - The stakeholder and the sponsor.
   - The project manager and the stakeholder.
   - The project manager and the sponsor.
   - The team and the project manager.

5. You are the project manager for a large government project. This project has a multi-million dollar budget, which will last 2 years and the contract was signed 6 months ago. You were not involved in contract negotiations or setting up procedures for managing changes, but now you are involved with changes from the customer and from people inside your organization. Who is normally responsible for formally reviewing major changes in the project or contract?
   - The change control board
   - The contracting/legal department
   - The project manager
   - Senior management

6. You are a new project manager for company (X). You previously worked for company (Y) that had an extensive project management practice. Company (X) has its own procedures, but you are more familiar with and trust those from company (Y).
   What should you do?
   - Use the practices from company (Y) but include any forms from company (X).
   - Use the forms from company (X) and begin to instruct them on ways to upgrade their own.
   - Talk about changes to the change control board of company (X).
   - Interact with others in an ethical way by sharing the good aspects of company (Y)'s procedures.

7. You, as the project manager, discover a defect in a deliverable that should be sent to the client under contract today. The project manager knows the client does not have the technical understanding to notice the defect. The deliverable technically meets the contract requirements, but it does not meet the project manager's fitness of standard use. What should the project manager do in this situation?
   - Issue the deliverable and get formal acceptance from the customer.
   - Note the problem in the lessons learned so future projects do not encounter the same problem.

- Discuss the issue with the customer.
- Inform the customer that the deliverable will be late.

8. Your customer requires a 3000 call capacity for the new call center project. However, one of your company's technical experts believes a 4000 call capacity can be reached. Another thinks that based on the technical needs of the customer, the capacity needs to be only 2500 calls. What is the BEST thing to do?
Meet with the customer to better understand the reasons behind the 3000 call capacity.
- Set the goal at 4000 calls.
- Meet with the technical experts and help them to agree on a goal.
- Set the goal at 3000 calls.

9. You are a project manager working on a multimillion-dollar project. As the project has progressed, you have become friends with the general contractor. You are working on a $100,000 change request. He has offered to let you use his villa on the coast for the next weekend as he will be away in another country. What should you do?
- Accept the offer with thanks.
- Decline the offer.
- Decline the offer and report it to your supervisor.
- Ask your boss to approve your use of the boat.

10. An employee approaches you and asks if he can tell you something in confidence. He advises you that he has been performing illegal activities within the company for the last year. He is feeling guilty about it and is telling you to receive advice as to what he should do. What should you do?
- Ask for full details.
- Confirm that the activity is really illegal.
- Inform your manager of the illegal activity.
- Tell the employee to inform their boss.

11. Near the end of the project, additional requirements were demanded by a group of stakeholders when they knew that they would be affected by your project. This became a problem because you had not included the time or cost in the project plan to perform these requirements. What is the learned lesson from this crisis?
- Review the WBS dictionary more thoroughly, looking for incomplete descriptions.

- Review the charter more thoroughly, examining the business case for "holes."
- Pay more attention to stakeholder management.
- Do a more thorough job of solicitation planning.

12. You are a project manager for a large installation project when you realize that there are over 100 potential stakeholders on the project. Which will be the best action?
    - Eliminate some stakeholders.
    - Contact your manager and ask which ones are more important.
    - Gather the needs of all the most influential stakeholders.
    - Find an effective way to gather the needs of all stakeholders.

13. What are the project management process requirements?
    - Initiating, developing, implementing, supporting
    - Initiating, planning, executing, controlling, closing
    - Feasibility, planning, design, implementation, supporting
    - Requirements analysis, design, coding, testing, installation, conversion, operation

14. One of your team members informs you that he does not know which of the many projects he is working on is the most important.
    Who should determine the priorities among projects in a company?
    - Project manager
    - Sponsor
    - Senior management
    - Team

15. You are trying to help project managers in your organization understand the project management process groups and the project management life cycle. Many of them are confusing the project life cycle with the project management life cycle. Which of the following identifies the DIFFERENCE between these two life cycles?
    - The project life cycle is created based on the top level of the work breakdown structure.
    - The project management life cycle is longer.
    - The project management life cycle only applies to some projects.
    - The project life cycle describes what you need to do to complete the work.

16. Which of the following statements BEST describes why stakeholders are necessary on a project?
    - They determine the project schedule, deliverables and requirements.
    - They help to determine the project constraints and product deliverables.
    - They supply the resources and resource constraints on the project.
    - They help provide assumptions, the WBS, and the management plan.

17. While working on a project in another country, you are asked to pay under the table money to facilitate the work to let the country officials issue a work order. What should you do?
    - Make the payment.
    - Ask the person for proof that the payment is required.
    - Seek legal advice on whether such a payment is a bribe.
    - Do not pay and see what happens.

18. You have just completed the design phase for a client's project and are about to enter the execution phase. All of the following need to be done EXCEPT:
    - Lessons learned
    - Updating records
    - Formal acceptance
    - Completion of the product of the project

19. Two days ago, you joined a consulting company as project manager to lead an existing project for a client. Today a major change is requested. What should be done FIRST?
    - Quickly develop a change control board to approve or disapprove changes.
    - Approve the change if your sponsor approved it, otherwise suggest a review by the project team.
    - Hire an outside consultant to develop and manage overall change control.
    - Find out if any formal overall change control plans and procedures are in place for this project.

20. The project has a critical deliverable that requires certain specialized and competent engineers to complete. The engineer who is working to complete the task has left the company and there is no one who

can complete the work within the company. For this reason, the project manager needs to acquire the services of a consultant as soon as possible.

What is the best action as the project manager?

- Follow the legal requirements set up by the company for using outside services.
- Bypass the company procedures as they are not relevant to the situation.
- Expedite and go directly to his/her preferred consultant.
- Ask his manager what to do.

21. You have just been hired as a project manager and you are trying to gain the cooperation of others.

What, in your opinion, is the BEST form of power for gaining cooperation under these circumstances?

- Formal
- Referent
- Penalty
- Expert

22. All of the following are correct statements about project managers EXCEPT:

- They are assigned after performance reports are distributed.
- They have the authority to say no when necessary.
- They manage changes and factors that create change.
- They are held accountable for project success or failure.

23. Which of the following is an output of team development?

- Management plan
- Staffing management plan
- Performance improvements
- Reward system

24. You have been working on a project for six months with the same team, yet the team still shows a lack of support for the project.

What should you do as a project manager to obtain the team's support for the project?

- Reevaluate the effectiveness of the reward system in place.
- Talk to each team member's boss with the team member present.
- Find someone else to be project manager.
- Tell the team he/she needs its support and ask them why they do not support the project.

25. The forms below present a power derived from the project manager EXCEPT:
    - Formal
    - Reward
    - Penalty
    - Expert

26. In which phase of the project should the project manager provide more direction?
    - Initiating
    - Planning
    - Executing
    - Controlling

2

Project Economic Analysis

2.1 Introduction

Assessing and managing investments in a new project involves the complex interaction of many variables.

This chapter explains all the variables that affect an economic study. Any big organization will have projects to perform, so to make decisions for the best project requires using economic tools and taking the risk assessment for each project into consideration.

Uncertainty is a parameter that affects an economic study, so the study needs to calculate the risk assessment of a project by using the Monte-Carlo simulation technique with the decision tree method. This method is most traditional in oil and gas projects and general industrial projects.

The major risks in any project will be the following:

- Economic risk and value
- Technical risk
- Political risk

Economic risk is affected by the market prediction, change of currency, inflation rate, oil price, and others. The technical risk is affected by the engineering study applying new technology, as it is possible to drill for wells and the wells be dry. Political risk depends on the country, as some countries have political stability and in other countries the political issues are unstable.

2.2   Project Cash Flow

Net cash flow is the key to all investment decisions, as it converts all elements of the project to the cost, and from that, one can compare different projects depending on the economic study.

The net cash flow (NCF) is used for the following reasons:

1. To measure the return of the project and liquidity over the work of the project
2. To calculate economic return by the net present value NPV
3. To calculate the risk assessment of the project
4. To reduce taxes on the life of the project

Net cash flow each year is calculated as the revenue from the project after subtracting the expense cost every year:

- Net cash flow = revenue − (operating cost + additional indirect expenses + taxes + investment + depreciation)

Revenue is the owner's income from the project every year depending on the volume of production that the project produces multiplied by the price for this product.

Operating cost consists of the direct cost, the cost of the materials used in the product, the indirect cost, the salary for the management level, computers, furniture, and others.

Taxes are a very critical item as they are the most time-consuming in developing NCF estimates. There are different types of taxes – production taxes, sales taxes, property taxes, state or region income taxes, and incorporate income taxes. So, these types of taxes and their value depend on the location of the project and the laws that govern it in the country that the project is located.

There is more than one way to compare different projects. A good rule of thumb is that any investment project should be profit-based. Therefore, the owner is normally involved in the feasibility study phase, making a comparison between more than one project in order to determine the revenue that suits the required interest rate that it deems appropriate to the company goal. There are different economic calculation methods that assist in decision-making and they are illustrated in this chapter.

Making a comparison is very important in feasibility studies for any project. It is important that in any of these methods you should identify the net cash flow.

Figure 2.1 shows the net cash flow diagram. For the beginning of the project, a lot of money will be spent to build the infrastructure or purchase machines and other necessary equipment required to deliver the required product. The value of these assets is called a capital cost (CAPEX) and it is usually spent only during the beginning of a project.

Assume that a project will start after one year. In this year you sold your product, so knowing the price and the number of products you will sell enables you to define the revenue in the first year and repeat it in the second year, third year, and for the lifetime of the project.

The number of products for every year can be known, but with uncertainty. Also, when you define the price of the product, this number is not easy to obtain as it varies from year to year and the uncertainty increases with time. Determining the value of a price requires a specialized consultant office for types of project investments, such as building hotels, which is different than a factory of kids' games, or rather, a steel factory. Therefore, strong market research is necessary for knowing the competitors in the market and the increase in the population of the country and a market

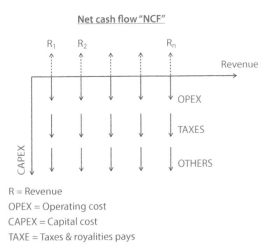

Figure 2.1 Net cash flow diagram.

research outside the country is helpful in the case of predicting whether to export the product.

In the oil and gas business, there are specialist teams in the main office of each international company that provide advice on calculations for each country. In addition to that, any country that has this development also has a team to perform this calculation, as the revenue, in this case, will be the volume of barrels of oil that can be produced every day which depends on the reservoir, where uncertainty lies in predicting its volume. However, the oil price will be defined on the head quarter as it needs a strategic expected plan for future prices.

For example, a reservoir for oil and gas can be predicted, but it constrains operational capabilities and managing the reserve to maintain the pressure inside the reservoir is limited by the production. Figure 2.2 shows that production will increase in the early few years, after that it will be stable for a few years, and then it will decline until reaching the uneconomic limit of the well, which should be defined. As this limit, the expenses will be higher than the gain, so the well will be shut off.

As shown in Figure 2.1, the revenue is a positive to the cash flow, but every year it will be an expense due to operation, maintenance to the equipment, and other expenses. Therefore, it will be a negative cash flow to the project. The operation and maintenance cost is called OPEX. In addition to the operation cost, there will be taxes that will be paid to the government, which will usually be a percentage from the production.

Any equipment has a lifetime. From an engineering point of view, the equipment and building lifetime is defined as being usable until it fails or is functionally not usable. But, there is also a financial lifetime from a financial point of view. All equipment has a value when it is new, but with time its value reduces, and when you need to sell, it will have another price. This price is called the book value. Each year there will be a depreciating cost

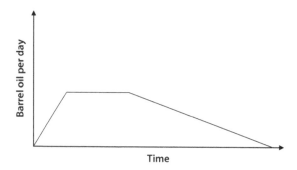

Figure 2.2 Production performance over project life time.

for this equipment and the value of this depreciation is related to financial rules for the government who will tax you.

Assume you purchase a new car that costs 20,000 dollars now. The next year you want to sell it and, according to the market, its price value is 15,000 dollars. To be easy in calculation, you assume an equation for reducing the money from $20,000 to $15,000. But, what about in 2014? So you return to the financial department to ask the book value in that year. Assume the book value is $18,000 in 2012, but in the market it will cost $15,000. This will be a problem. In a private company this is not a problem, as you will be the loser. But for a government or public company, this will be a problem, especially in developing countries where the rules are very restricted, requiring that you sell the car based on the book value or higher.

There is usually a financial argument between the company's financial department and the government's taxes department as the depreciating value of the equipment will be deducted from the taxes. So the rate of depreciation is calculated by different methods, but any of these methods are used based on the government's taxes, rules, and laws. The all common depreciation method will be illustrated.

2.2.1   Depreciation Methods

Capitalized items to depreciation include casing, tubing, flow lines, tanks, platforms, and the like. Tangible investments are basically related to any item that has a life in excess of one year.

Campbell, *et al.* (1987) mention that the depreciation write-offs depend on the type of expenditures. The law presently divides investment into two categories: five- and seven-year lives. The depreciation schedule for each project life is presented. The life year values under the 1987 laws in the USA are derived from a 200 percent declining balance over five years, often referred to as a 40 percent decline balance in Europe, and in Canada depreciation in the seven-year category converts to a 28.6 percent declining balance in this area. The seven-year category covers virtually every petroleum investment except for drilling equipment (five years), oil refining equipment (ten years), and transmission pipeline and related equipment (fifteen years).

Depreciation begins when the project is classified as "ready for service." The definition of ready of service varies among companies. Some define ready for service as capable of producing, while others require production to actually take place before depreciation. Gas wells, for example, are often completed but not hooked up. Process plants take several years to

Table 2.1 Financial life time.

5 year life	Automobiles
	Trucks
	Drilling equipment
	Computer
10 year life	Crude oil refining equipment
15 year life	Transmission pipeline and equipment
31.5 year life	Buildings

build, start-up offshore platforms are constructed, and wells are drilled, but production cannot commence until the product is disposed of. The same situation exists in remote, onshore areas. Each example experiences a delay between the completion of the project from a technical perspective and the initial flow of revenue.

The question is whether a tangible investment that is capable of working, but not active, is ready for service and can be depreciated. Opinions and practice vary. Some argue that capable of working is the same as ready for service. Others take a more conservative position and begin depreciation only when sale begins. Decisions regarding the correct approach depend on the legal staff's opinion about what they can justify in court. The capable of working interpretation lowers the after tax cost of investment by beginning depreciation writes-off earlier. The various depreciation methods will be as follows for different industrial projects:

2.2.1.1   Straight-Line Method

The rate of depreciation is calculated by the following equation:

$$D_t = (1/n)\,(C{-}SV), \tag{2.1}$$

where C equals original cost of capitalized investment, D_t equals depreciation in year, t, SV equals salvage value of capitalized investment, and N equals number of years of depreciation.

This method can be illustrated by the following example, where it is assumed that the value of the equipment after five years is equal to $10,000 and this price is based on the value of the equipment at the sale after five years.

It is worth mentioning that salvage value is usually taken to equal zero.

Depreciation rate per year = (50000–10000)/5 = $8,000.

2.2.1.2   Declining-Balance Method

$$D_t = 1 - \left(\frac{sv}{c}\right)^{\frac{1}{4}}. \tag{2.2}$$

Depreciation rate per year $= 1 - \left(\dfrac{10000}{50000}\right)^{1/5}.$

2.2.1.3   Sum-of-the-Year-Digits

$$D_t = (y/SYD)(C-SV), \tag{2.3}$$

where Y equals the number of years remaining in depreciation and SYD equals the sum of the year digits.

Table 2.2 Straight-line method.

Book value at the end of the year	Depreciation value per year	Year
50000	8000	0
42000	8000	1
34000	8000	2
26000	8000	3
18000	8000	4
10000	8000	5

Table 2.3 Declining–balance method.

Book value at the end of the year	Depreciation value per year	Year
50000		0
35000	50000 × 0.3 = 15000	1
24500	35000 × 0.3 = 10500	2
17150	24500 × 0.3 = 7350	3
12000	17150 × 0.3 = 5150	4
8400	12000 × 0.3 = 3600	5

Table 2.4 Sum-of-the-year-digits.

Book value at the end of the year	Depreciation value per year	Year
50000		0
36700	$40000 \times (5/15) = 13333$	1
26000	$40000 \times (4/15) = 10670$	2
18000	$40000 \times (3/15) = 8000$	3
12600	$40000 \times (2/15) = 5330$	4
10000	$40000 \times (1/15) = 2670$	5

Table 2.5 Sinking-fund method.

Book value at the end of the year	Depreciation value per year	Year
50000	–	0
42900	7100	1
35380	7520	2
27410	7970	3
18960	8450	4
10000	8960	5

2.2.1.4   Sinking-Fund Method

$$D_t = (C-SV)*F \qquad (2.4)$$

where F equals the factor for many values at five years for an interest rate assuming seven percent.

First year = $(50000 - 10000)(0.1774) = 7069$.
Second Year = $7096 + 0.06(7096) = 7521.8$.

2.2.1.5   Service-Out Method

This method depends on the time period in which the equipment operates, for example, drilling equipment or digging tunnels where the rate of depreciation is associated with use. Assume that the cost of the equipment is around $12,0000 and the value of the equipment when sold at the end of the project, which is at the dig about 15,0000 meters, is $6000. In this case, we calculate the depreciation rate from the following equation:

Depreciation rate = $(120000-6000)/150000 = 0.76$ $/m.

Therefore, it is used to calculate the value of the equipment after the first year, as it is known that the equipment was used to drill a total depth equal to 30,000 meters, as shown from the following equation:

Value of the equipment = 120000 – (0.76 * 30000) = $97,200.

2.2.2   Method of Net Present Value (NPV)

The method of Net Present Value (NPV) is one of the most common, widely used methods and is one of the most important ways upon which the decision-making tree will be displayed later.

It is important to identify the lifetime of the project and the distribution of cash flow during years of the project from its beginning to the end of the project's lifetime.

To calculate the equation by the current value, one must specify the discount rate which is given by the following equation:

$$NPV = \frac{NCF_1}{(1+D)^1} + \frac{NCF_2}{(1+D)^2} + \ldots + \frac{NCF_n}{(1+D)^n}, \quad (2.5)$$

where n is the number of years and D is the discount rate that is equal to the value of the interest rate.

The following example illustrates the NPV calculation in the suit in which the imposition of the discount rate is equal to ten percent.

As an example, calculating the net present value is shown in Table 2.6.

2.2.2.1   Inflation Rate

When assuming the interest rate or discount rate, one must not forget the inflation rate. The rate of inflation may be assumed as a variable every year or a fixed value and there are studies and economic research to calculate

Table 2.6 Net present value calculation.

Net present value	Net cash flow	Discount rate	Year
−51785.00	−51785	1.0	0
18181.80	20000	0.909	1
16528.80	20000	0.826	2
0015026.20	20000	0.7513	3
13659.00	20000	0.683	4
11610.80		NPV	

the rate of inflation. Its value differs from one country to another according to a country's nature of economics. The following equation is calculated using the highest rate of return and includes rate of inflation:

The nominal interest rate of return = (1 + inflation rate)
(1 + interest rate) −1.   (2.6)

The previous example includes an interest rate of ten percent and assumes an inflation rate of four percent. The real interest rate or discount rate is figured by the following equation:

$$D = \frac{1.10}{1.04} = 1.0577.$$ (2.7)

When the inflation rate is constant, the net present value is stable under the assumption that the rate of return is fixed. When the difference in the value of the interest rate changes every year, the net present value will be different.

2.2.3   Minimum Internal Rate of Return (MIRR)

This method depends on calculating the value of D. Let the NPV equal zero so that it will be calculated by the trial and error method.

$$NPV = NCF_1(1+D)^{-1} + NCF_2(1+D)^{-2} + \ldots\ldots + NCF_n(1+D)^{-n}$$ (2.8)

By knowing the NCF and that the imposition of NPV is equal to zero, the value of D is the project interest rate of return that the company, organization, or individual investor will achieve after implementing the project.

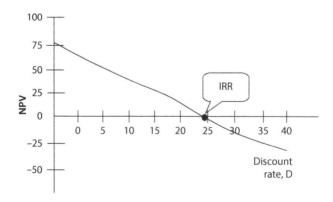

Figure 2.3 Calculate IRR.

Every company should have defined their own minimum rate of return (MIRR). This rate is internal and specific to each company. For most international petroleum companies with branches in more than one country around the world, that number varies from country to country.

This number is determined after studies and much research specific to each country according to the description of the political, social, and economic condition of that country. For example, investing in England or the USA is certainly different from Sudan or Nigeria and other African countries that do not have any political stability.

The internal minimum of return number is a secret and confidential number for each company as these numbers govern their investment.

2.2.4   Payout Method

This method is the fastest and easiest way to calculate the time required to recover the invested money in the project, but it cannot account for the project interest rate of return. Calculating the time period is simple, as shown in Figure 2.4.

The time required to recover the money invested is called payout time and is also known as a time period equal to the cost of the project with the return of the project. This factor depends on the expertise of the decision-maker, as this is a very important factor in a country with no political stability and requires that the project decision-maker be able to recover the capital in the shortest possible time.

The disadvantage of this method is that it cannot calculate the interest rate of return for a project, which varies according to each project and the

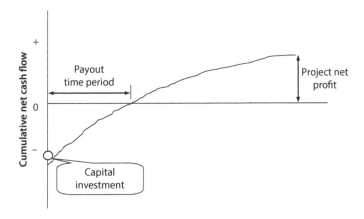

Figure 2.4 Payout method.

value of the initial investment. But, in general, given a general idea of the itinerary of the project economically and, as mentioned in some invest-ments, may be a time period of recovery of funds that was paid for the cost of the project and is the most important factor affecting the decision-making regardless of the final profit.

In spite of the many advantages of the payout method, alone it is not a complete measure of the value of money for the following reasons:

- It does not indicate profit following payout.
- It does not measure total profit.
- Time value of money is not formally included.
- It varies depending on different types and magnitudes of investment.

2.3   Economic Risk Assessment

2.3.1   Probability Theory

To enter into the theory of probability, there is some statistical informa-tion that is important and necessary to understand the probability theory and the probability distribution. To illustrate the statistical concepts, we will clarify them through a numerical analysis for test results for crushing samples of cylinder concrete to measure its strength.

The main statistical parameters will be the following:

- Arithmetic average
- Standard deviation
- Coefficient of variation

Arithmetic average is the average value of a set of results and is repre-sented in the following equation:

$$\overline{X} = \frac{X_1 + X_2 + \ldots + X_n}{n},\qquad(2.9)$$

where n is the number of results and X is read of each test result.

As a practical example of the statistical parameter, assume that we have two groups of concrete mixture from different ready mix suppliers. The first group has a concrete compressive strength after 28 days for three samples which are 310 kg/cm^2, 300 kg/cm^2, and 290 kg/cm^2. When we cal-culate the arithmetic average using Equation 2.9, the arithmetic mean of these readings is 300 kg/cm^2.

The second group has a test result for cube compressive strength after 28 days under the same conditions for the first group. The test results are 400 kg/cm², 300 kg/cm², and 200 kg/cm². When calculating the arithmetic mean, we find that it is equal to 300 kg/cm².

Because the two groups have the same value of the arithmetic mean, does that mean that the same mixing has the same quality? Will you accept the two mixing? We find that this is unacceptable by engineering standards, but when we consider that the mean of the two groups are the same, one should choose another criteria by which to compare the results as we cannot accept the second group based on our judgment, which will not support us in court.

Standard deviation is a statistical factor that reflects near or far the reading results. From the arithmetic mean end, it is represented in the following equation:

$$S = \sqrt{\frac{(X_1 - \overline{X})^2 + (X_1 - \overline{X})^2 + \ldots\ldots + (X_n - \overline{X})^2}{n}}. \qquad (2.10)$$

The standard deviation for the first group sample is

$$S = \sqrt{\frac{(310 - 300)^2 + (300 - 300)^2 + (290 - 300)^2}{3}}.$$

Mixing one, $S = 8.16$ kg/cm².
The standard deviation for the seconding group sample is

$$S = \sqrt{\frac{(400 - 300)^2 + (300 - 300)^2 + (200 - 300)^2}{3}}.$$

Mixing two, $S = 81.6$ kg/cm²

One can find that the standard deviation in the second group has a higher value than the first group. So the distribution of test data results is far away from the arithmetic mean rather than group one. From Equation 2.10, one can find that the ideal case is when S equals 0.

We note that the standard deviation has units, as seen in the previous example. Therefore, standard deviation can be used to compare between the two groups of data as in the previous example where the two groups give the value of 300 kg/cm² after 28 days. On the other hand, in the case of the comparison between the two different mixes of concrete, for instance, there is a resistance of 300 kg/cm² in one concrete and 500 kg/cm² in the second. In that case, the standard deviation is of no value. Therefore, we resort to the coefficient of variation.

The coefficient of variation is the true measure of quality control, as it determines the proportion after the readings for the average arithmetic profile. This factor has no units and is, therefore, used to determine the degree of product quality.

$$C.O.V = \frac{S}{\overline{X}} \qquad (2.11)$$

As another example, assume there is a third concrete mix at another site to provide concrete strength after 28 days of 500 kg/cm². When you take three samples, it gives the results of strength after 28 days as 510 kg/cm², 500 kg/cm², and 490 kg/cm². When calculating the arithmetic mean and standard deviation, we find the following results:

- Arithmetic mean = 500 kg/cm²
- Standard deviation = 8.16 kg/cm²

So, when comparing between concrete from one site with a mean concrete strength of 300 kg/cm² and concrete from a second site with a mean concrete strength of 500 kg/cm² and the two sites have the same standard deviation as the above example, the coefficients of variations are as follows:

- Coefficient of variation of the first site = 0.03
- Coefficient of variation of the second site = 0.02

We note that the second site has a coefficient of variation less than the first location. That is, the standard deviation to the arithmetic average is less at the second site than at the first site. This means that the second site mix concrete has a higher quality. Therefore, the coefficient of variation is the standard quality control of concrete and the closer to zero, the better the quality control.

To define a practical method of probability distribution, assume we have test results for 46 cube crushing strengths, as shown in Table 2.7, and we need to define the statistical parameters for these numbers.

The raw data is collected in groups and the number of samples with a value between the range for every group is called frequency. The frequency table from the raw data in Table 2.8 is tabulated in Table 2.9.

The data from Table 2.9 is presented graphically in Figure 2.5. To analyze the data, make a cumulative descending table, as shown in Table 2.8. This table is presented graphically in Figure 2.6. From Table 2.10, one can find that for this set of data, sample results of concrete strength can obtain the cumulative descending data. To understand this table, the probability of having a concrete strength that is less than 300kg/cm² is nine percent.

Table 2.7 Net present value including inflation.

Year	(1) Net cash flow	(2) Inflation rate	(3) = (1) × (2) Net cash flow after inflation	(4) Discount rate	(5) = (4) × (3) Net present value
0	−51785	1.0	−51785	1.0	−51785
1	20000	0.96	19231	0.95	18182
2	20000	0.92	18491	0.89	16529
3	20000	0.89	17780	0.85	15026
4	20000	0.85	17096	0.80	13660
Sum (NPV)					11612

Table 2.8 Row data.

340	298	422	340	305
356	320	382	297	267
355	312	340	366	349
311	306	368	382	404
326	350	322	448	350
358	384	346	365	303
398	306	298	339	344
378	282	320	360	360
367	341	326	325	352
384				

Table 2.9 Frequency table.

Frequency	Average value	Group	ID
1	270	260–280	1
3	290	280–300	2
6	310	300–320	3
6	330	320–340	4
12	350	340–360	5
7	370	360–380	6
5	390	380–400	7
1	410	400–420	8
1	430	420–440	9
1	450	440–460	10
43	Total		

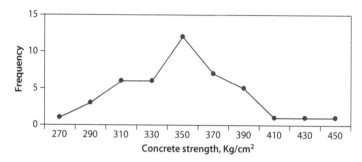

Figure 2.5 Frequency curve for concrete compressive strength data.

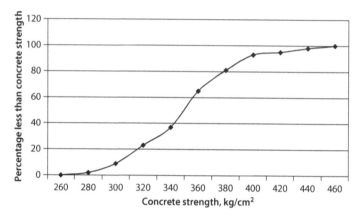

Figure 2.6 Cumulative distribution curve for concrete strength.

Table 2.10 Descending cumulative table.

Group no.	Test value	Reading value less than the upper limit	The percentage less than the upper limit
10	460	43	100
10	440	42	98
9	420	41	95
8	400	40	93
7	380	35	81
6	360	28	65
5	340	16	37
4	320	10	23
3	300	4	9
2	280	1	2
1	260	0	0

From the cumulative descending curve, one can find that 100 percent of the results of the samples have a strength less than 459 kg/cm² at the same time. In the results of previous tests, we find that the samples have results less than or equal to 280 kg/cm², which is about two percent of the number of tested samples.

2.3.2   Probability Distribution of Variables

Most civil engineering problems deal with quantitative measures in the familiar deterministic formulations of engineering problems. However, there is nothing deterministic at all, when you assume you have a reinforced concrete column in drawings mentioning that its section is 500 millimeters by 500 millimeters. This means it will be this exact number when you measure the column. It is possible that there could be some deviation, which is allowable in the code. So, the column section dimensions are not deterministic.

The concept of mathematical variables and functions of variables have proven to be useful substitutes for less precise qualitative characteristics. Variables, whose specific values cannot be predicted with certainty before an experiment, can be presented by the probabilistic models and distributions.

2.3.2.1   Normal Distribution

Normal distribution is used to represent many natural phenomena, such as the lengths of people, and it is used in decision-making as it can present the inflation rate or the price of oil in the future. This distribution is widely used in metering equipment, as it represents the measurement error and the permeability of the soil and the spaces between the grains and saturation as well as some economic data.

Equation:

$$f(x) = \left(\frac{1}{\sigma 2\pi^{0.5}} \right) e^{-0.5(x-\mu)^2/\sigma^2} \qquad (2.12)$$

Mean:

$$\bar{x} = \frac{\sum x_i}{n} \qquad (2.13)$$

where:
$\bar{x}$ = arithmetic mean of sample data
x_i = each individual value in sample

n = number of values in sample
cm = class mark
nc = number of values in class,
Standard deviation is given by

$$\sigma_s = \left[\left(\frac{1}{n}\right)\sum x^2 - \mu^2\right]^{0.5} \tag{2.14}$$

where σ is the standard deviation and μ is the arithmetic mean.

Normal distribution is the most commonly used probability distribution because it was found identical with the most natural phenomena. It was found that normal distribution is the best probability curve to present concrete strength from laboratory tests performed on the concrete in most countries of the world to present the concrete.

The characteristics of this distribution are as follows:

- Normal distribution is the distribution symmetrically around the average and, more precisely, the arithmetic mean of the curve is divided into two equal halves
- Normal distribution matches the arithmetic mean and median lines and mode value to find the most likely to occur.

The area under the curve is equal to one and the random variable as a result of concrete cube strength, for example, can take the values from ∞ to $-\infty$. So, this curve presents all the possible values of concrete strength.

As a result, each curve depends on the value of the arithmetic mean and standard deviation and any difference between the two parameters leads to a difference in the shape of the probability distribution. Therefore, the standard normal distribution is used to determine areas under a curve by knowing the standard deviation and arithmetic mean. Another variable, z, is obtained from the following equation:

$$z = \frac{x - \overline{x}}{\sigma} \tag{2.15}$$

Table 2.2 shows the values of the area under the curve by knowing the value of z from the above equation.

In the first column, the value of z and first row determine the accuracy to the nearest two decimal digits. From the table, one can find that the area under the curve at z is equal to 1.64 is 0.4495 and the area under the curve for any values less than z is 0.5 to 0.4495, which is equal to about 0.0505. In other words, the probability of the variables has a value less than equal to and less than five percent, as shown in Figure 2.7.

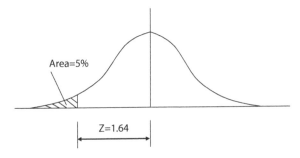

Figure 2.7 Normal distribution curve.

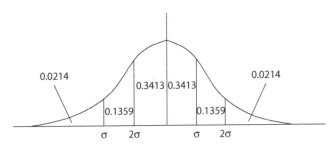

Figure 2.8 The division of areas under the normal distribution.

Figure 2.8 shows the area under the curve when you add or decrease the value of the standard deviation of the arithmetic mean.

We note from the previous format that the area under the curve from the arithmetic mean value to one value of the standard deviation is equal to 34.13 percent, while in the area under the curve in Figure 2.8, for twice the standard deviation, the value is equal to 47.72 percent.

2.3.2.2   Log Normal Distribution

This distribution is used when a phenomenon does not take a negative value. Therefore, it is used in representing the size of the aquifer or reservoir properties, such as permeability of the soil. It also represents real estate prices and others.

Equation:

$$f(x) = \left(\frac{1}{2\pi.\beta}\right) x^{-1} e^{-(inx - \sigma^2/2\beta^2)} \tag{2.16}$$

Mean:

$$\ln \bar{x}_G = \frac{\sum (\log x)}{n} = \mu_G \tag{2.17}$$

Table 2.11 The area under the curve of normal distribution.

Z	0	0.01	0.02	0.03	0.04	0.05	0.06	0.07	0.08	0.09
0	0	0.004	0.008	0.012	0.016	0.0199	0.0239	0.0279	0.0319	0.0359
0.1	0.0398	0.0438	0.0478	0.0517	0.0557	0.0596	0.0636	0.0675	0.0714	0.0753
0.2	0.0793	0.0832	0.0871	0.091	0.0948	0.0987	0.1026	0.1064	0.1103	0.1141
0.3	0.1179	0.1217	0.1255	0.1293	0.1331	0.1368	0.1406	0.1443	0.148	0.1517
0.4	0.1554	0.1591	0.1628	0.1664	0.17	0.1736	0.1772	0.1808	0.1844	0.1879
0.5	0.1915	0.195	0.1985	0.2019	0.2054	0.2088	0.2123	0.2157	0.219	0.2224
0.6	0.2257	0.2291	0.2324	0.2357	0.2389	0.2422	0.2454	0.2486	0.2517	0.2549
0.7	0.258	0.2611	0.2642	0.2673	0.2704	0.2734	0.2764	0.2794	0.2823	0.2852
0.8	0.2881	0.291	0.2939	0.2967	0.2995	0.3023	0.3051	0.3078	0.3106	0.3133
0.9	0.3159	0.3186	0.3212	0.3238	0.3264	0.3289	0.3315	0.334	0.3365	0.3389
1	0.3413	0.3438	0.3461	0.3485	0.3508	0.3531	0.3554	0.3577	0.3599	0.3621
1.1	0.3643	0.3665	0.3686	0.3708	0.3729	0.3749	0.377	0.379	0.381	0.383
1.2	0.3849	0.3869	0.3888	0.3907	0.3925	0.3944	0.3962	0.398	0.3997	0.4015
1.3	0.4032	0.4049	0.4066	0.4082	0.4099	0.4115	0.4131	0.4147	0.4162	0.4177
1.4	0.4192	0.4207	0.4222	0.4236	0.4251	0.4265	0.4279	0.4292	0.4306	0.4319
1.5	0.4332	0.4345	0.4357	0.437	0.4382	0.4394	0.4406	0.4418	0.4429	0.4441
1.6	0.4452	0.4463	0.4474	0.4484	0.4495	0.4505	0.4515	0.4525	0.4535	0.4545
1.7	0.4554	0.4564	0.4573	0.4582	0.4591	0.4599	0.4608	0.4616	0.4625	0.4633
1.8	0.4641	0.4649	0.4656	0.4664	0.4671	0.4678	0.4686	0.4693	0.4699	0.4706

1.9	0.4713	0.4719	0.4726	0.4732	0.4738	0.4744	0.475	0.4756	0.4761	0.4767
2	0.4772	0.4778	0.4783	0.4788	0.4793	0.4798	0.4803	0.4808	0.4812	0.4817
2.1	0.4821	0.4826	0.483	0.4834	0.4838	0.4842	0.4846	0.485	0.4854	0.4857
2.2	0.4861	0.4864	0.4868	0.4871	0.4875	0.4878	0.4881	0.4884	0.4887	0.489
2.3	0.4893	0.4896	0.4898	0.4901	0.4904	0.4906	0.4909	0.4911	0.4913	0.4916
2.4	0.4918	0.492	0.4922	0.4925	0.4927	0.4929	0.4931	0.4932	0.4934	0.4936
2.5	0.4938	0.494	0.4941	0.4943	0.4945	0.4946	0.4948	0.4949	0.4951	0.4952
2.6	0.4953	0.4955	0.4956	0.4957	0.4959	0.496	0.4961	0.4962	0.4963	0.4964
2.7	0.4965	0.4966	0.4967	0.4968	0.4969	0.497	0.4971	0.4972	0.4973	0.4974
2.8	0.4974	0.4975	0.4976	0.4977	0.4977	0.4978	0.4979	0.4979	0.498	0.4981
2.9	0.4981	0.4982	0.4982	0.4983	0.4984	0.4984	0.4985	0.4985	0.4986	0.4986
3	0.4987	0.4987	0.4987	0.4988	0.4988	0.4989	0.4989	0.4989	0.499	0.499
3.1	0.499	0.4991	0.4991	0.4991	0.4991	0.4992	0.4992	0.4992	0.4992	0.4993
3.2	0.4993	0.4993	0.4994	0.4994	0.4994	0.4994	0.4994	0.4994	0.4995	0.4995
3.3	0.4995	0.4995	0.4995	0.4996	0.4996	0.4996	0.4996	0.4996	0.4996	0.4997
3.4	0.4997	0.4997	0.4997	0.4997	0.4997	0.4997	0.4997	0.4997	0.4997	0.4998
3.5	0.4998	0.4998	0.4998	0.4998	0.4998	0.4998	0.4998	0.4998	0.4998	0.4998
4.0	0.49997									
5.0	0.49999									

Standard Deviation:

$$\ln \sigma_s = \left[\left(\frac{1}{n} \right) \Sigma (\ln x)^2 - (\ln \mu_G)^2 \right]^{0.5}$$

(2.18)

2.3.2.3   Binominal Distribution

This distribution is used for the following reasons:

- To determine geological hazards
- To calculate the performance of the machine for the cost and the cost of spare parts
- To determine the appropriate number of pumps with the appropriate pipeline size with the required fluid capacity and the number of additional machines.
- To determine the number of generators according to the requirement of the project and to determine the number of additional generators in the case of an emergency or malfunction in any machine.

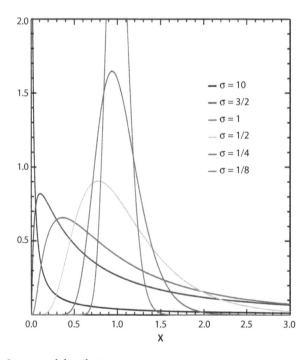

Figure 2.9 Lognormal distribution.

To understand the nature of this distribution let us use the following:
Equation:

$$f(x) = \frac{n!}{n_s!(n-n_s)!}(f)^{n_s}(1-f)^{n-n_s}$$
(2.19)

Mean:

$$x = n.f$$
(2.20)

Standard Deviation:

$$\sigma_s = \left[n.f.(1-f)\right]^{0.5}$$

Example 1:

When playing by the coin, the probability of the queen appearing is $P = .50$. What is the probability that we get the queen twice when we lay down the currency 8 times?

$$F(x) = [8!/2!(6!)]\,(0.5)^2\,(0.5)^{8-2}$$
$$= 0.189$$

This means that when you take a coin 6 times, the probability that the image will appear twice is 0.189.

Example 2:

Assuming the probability of 0.7 when drilling a single well that has oil, what is the probability that we find oil in 25 wells when we drill 30 wells?

Therefore, we find that the likelihood of success of the individual well is 0.7, but the possibility that the 25 successful wells were drilled is 0.0464.

Example 3:

Assess the reliability of a system requiring 10,000 KW to meet system demand. Each generator has been rated 95% reliable (5% failure rate).

The company is comparing 3 alternatives: 2–5000 KW generators, 3–5000 KW, and 3–4000 KW generators.

When we do a comparison between normal and logarithmic distributions and the binominal distribution and look at the shape of each of the three curves, we find that the log and normal distribution curves are solid curves which are different than the binominal distribution curve,

Table 2.12 Alternative for Example 3.

2–5000		3–5000	3–4000	
10,000	0.9025	0.9928	12,000	0.8574
5,000	0.0950	0.0071	8,000	0.1354
0	0.0025	0.0001	4,000	0.0071
			0	0.0001
Total	1.000	1.000		1.000
Avg. Reliability	0.9500	0.9963		0.9685

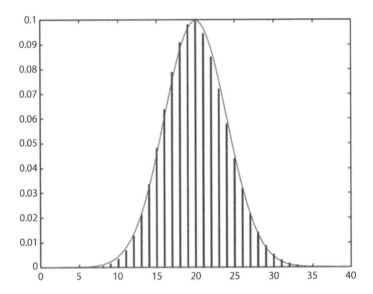

Figure 2.10 Binominal distribution.

as in Figure 2.10. The curve in Figure 2.10 presents by rectangular bars. Therefore, the normal and lognormal distribution is called the Probability Density Function (PDF). This PDF distribution curves are used in cases of descriptions of natural phenomenon or material that can take any figure, for example, when you calculate the lengths of people in the building that you are in. You will find that the lowest number, for example, is 120.5 centimeters and the largest number is 180.4 centimeters and the lengths of people can be any number between those numbers. But in the case of the last example, the number of drilling wells is between one and twenty-five wells in calculating the probability of success at a specific number of wells. So, we calculate the probability of success for twenty wells and cannot say that the possibility of drilling wells 20.511. Therefore, in that case, this probability distribution will be called the Probability Mass Function (PMF). This is very important when choosing the suitable distribution, which should match the natural phenomena for these variables. When defining the probability distributions for steel strength, oil price, or population, one should use the probability density function (PDF).

2.3.2.4 Poisson Distribution

This distribution is based on the number of times the event occurs within a specific time period, such as the number of times the phone rings per minute or the number of errors per page of a document overall and that description is used in transport studies or in deciding upon the number of fuel stations to fuel cars, as well as in the design study for telephone lines.

Mean:

$$m_t = \lambda \tag{2.22}$$

Standard deviation:

$$\sigma = \lambda \tag{2.23}$$

It will be a probabilistic mass function, as shown in Figure 2.11.

2.3.2.5 Exponential Distribution

This distribution represents the time period between the occurrences of random events. For example, the time period between the occurrences of electronic failures in equipment reflects this distribution and is the opposite of Poisson distribution. It is used in the time period that occurs in machine failures and there are now extensive studies that use this model to determine the appropriate time period for maintenance of equipment, called mean time between failure (MTBF).

Probability Density Function:

$$f_T(t) = \lambda e^{-\lambda t} \tag{2.24}$$

Mean:

$$M_t = 1/\lambda \tag{2.25}$$

Standard deviation:

$$\sigma = 1/\lambda \tag{2.26}$$

2.3.2.6 Weibull Distribution (Rayleigh Distribution)

Wind speed is one of the natural phenomena for which we use the Weibull distribution. It is also used to stress test metals and to study quality control

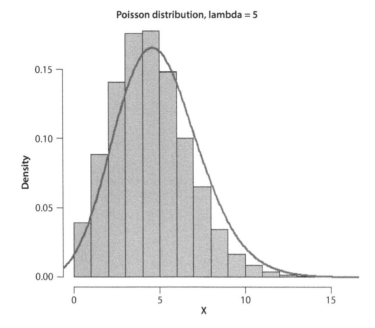

Figure 2.11 The exponential distribution.

or machines reliability and the time of the collapse. This distribution is complicated and, therefore, is not recommended for use in the case of building a huge model of an entire problem when using the Monte-Carlo simulation.

2.3.2.7   Gamma Distribution

This distribution represents a large number of events and transactions, such as inventory control or representation of economic theories. The theory of risk insurance is also used in environmental studies when there is a concentration of pollution. It is also used in studies where there is petroleum crude oil and gas condensate and it can be used in the form of treatment in the case of oil in an aquifer.

Equation:

$$f(x) = \frac{1}{\beta^{a} T(a)} x^{a-1} e^{-x/\beta}$$

(2.27)

Mean:

$$x = a\beta$$

(2.28)

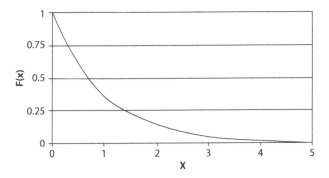

Figure 2.12 The exponential distribution.

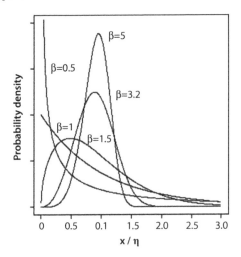

Figure 2.13 Weibull Distribution.

Standard Equation:

$$\sigma_s = \sqrt{a\beta^2}$$

(2.29)

The different shapes of gamma distribution are presented in Figure 2.14.

2.3.2.8 Logistic Distribution

This distribution is used frequently to describe the population growth rate in a given period of time. It can also represent interactions between chemicals.

$$f(x;\mu,s) = \frac{e^{-(x-\mu)/s}}{s\left(1+e^{-(x-\mu)/s}\right)^2}$$

(2.30)

2.3.2.9 Extreme Value (Gumbel Distribution)

This distribution is used when the intended expression of the maximum value of the event occurs in a period of time. Therefore, it is used for floods, earthquakes, or rain and is used to calculate the loads on the plane and study the fracture resistance of some materials.

$$f_z(z) = a\exp\left[a(z-u) - e^{a(z-u)}\right]$$ (2.31)

$$F_z(z) = 1 - \exp(-e^{a(z-u)}),$$ (2.32)

where $-\infty \le z \le \infty$.

$$m_z = u - \frac{\gamma}{a}.$$ (2.33)

$$\sigma_z = \frac{\pi}{a\sqrt{6}}.$$ (2.34)

2.3.2.10 Pareto Distribution

This distribution is commonly used when characterizing per capita income, a change in stock price, the size of the population in a city, the number of staff in a company, as well as the errors that occur in a communications circuits. It also represents changes in natural resources.

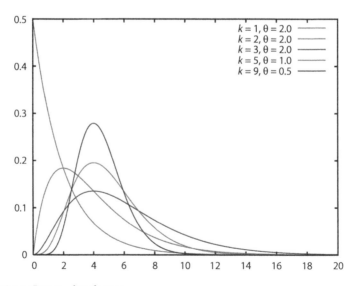

Figure 2.14 Gamma distribution.

It uses the Pareto principal, which is based on the idea that by doing twenty percent of the work, 80 percent of the advantage of doing the entire job can be generated. Or, in terms of quality improvement, a large majority of problems (80 percent) is produced by a few key causes (20 percent).

2.3.3   Distribution for Uncertainty Parameters

There are some variables that are difficult to identify in distributions, such as the project estimate cost or the experiments and tests necessary to study the phenomenon that are very expensive, such as defining the area and height of an oil reservoir, so we use the following distributions.

2.3.3.1   Triangular Distribution

This distribution is very important in the case of phenomenon where testing is very expensive. An example is when you select the size of an underground reservoir and three tests are usually performed to obtain the minimum, the maximum, and most likely. This distribution is used in schedule planning, which will be discussed in Chapter 4, and it determines the time required to complete the activity by three values, a minimum time, maximum time, and the most likely time to finish that activity.

In addition, it is also used to determine the estimated cost of a project where the maximum allowable value is about a ten to fifteen percent increase on the cost and the calculated minimum value is a ten to fifteen percent decrease for the calculated cost.

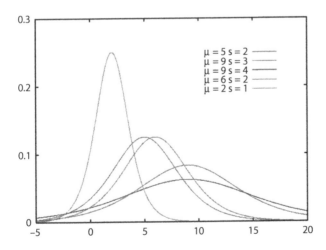

Figure 2.15 Logistic distribution.

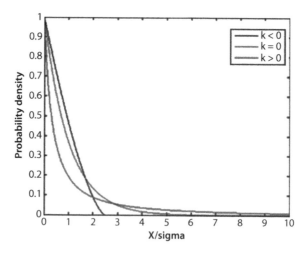

Figure 2.16 Pareto distribution.

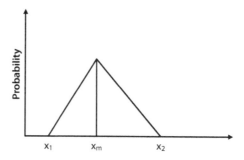

Figure 2.17 Triangular distribution.

Triangular distribution is shown in Figure 2.17 where x_1, x_2, and x_m are the minimum, maximum, and most likely values, respectively.

Equation:

$$x = \frac{x_1 + x_m + x_2}{3} \tag{2.35}$$

$$\sigma_s = \left[\frac{\left(x_2 - x_1\right)\left(x_2^2 - x_1 x_2 + x_1^2\right) - x_m x_2 \left(x_2 - x_m\right) - x_1 x_m \left(x_m - x_1\right)}{18\left(x_2 - x_1\right)} \right]^{0.5} \tag{2.36}$$

2.3.3.2   Uniform Distribution

This distribution is shown in Figure 2.18 and is used in the case that the event can take place at any value with the same probability of occurrence.

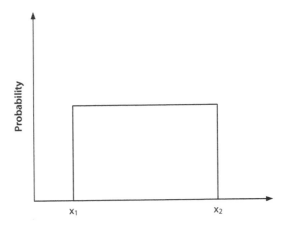

Figure 2.18 Rectangular Distribution.

For example, when you roll a die the probability of rolling a number between one and six is the probability of a constant, which is the probability of 1/6.

x_1 and x_2 are the minimum and maximum values, respectively.

$$x = \frac{x_1 + x_2}{2} \tag{2.37}$$

$$\sigma_s = \left[\frac{\left(x_1 - x_2\right)^2}{12} \right]^{0.5} \tag{2.38}$$

2.3.4   Choose the Appropriate Probability Distribution

Each type of probability distribution has its own properties, which gives every distribution the ability to represent a specific natural phenomenon. For example, we find that normal distribution is a good distribution that can represent concrete strength and an increase in population can be represented by logistic distribution. So, before building a model, one must be sure to choose the best probability distribution that represents the parameter.

One can obtain the suitable probability distribution by returning to previous references or research, as many statistical studies have been performed for most engineering parameters. Another way to obtain the suitable probability distribution is by studying the phenomenon to define the best distribution to represent it. The second is performed through a test more than once, and the results are plotted and compared to the probability distributions. The most common methods used to choose the best probability distributions that match with the phenomena test results are the K-S and Chi square methods.

As mentioned earlier, there is a way to choose the appropriate distribution mathematically, but each distribution has certain properties.

If there is raw data, as in the example of concrete strength, do the same procedure to define the frequency tables and curve by trying to choose the best probability distribution that can match with this curve.

2.3.4.1   Chai Square Method

This method calculates the potential value obtained from the practical test (O) and we calculate the corresponding value when calculated from the mathematical equation for the probability distribution (E). We then apply the following equation where (χ) is what we repeat in calculating its value according to the number of groups identified previously, as in the previous example, ten times, so k equals 10 in the following equation:

$$\chi^2 = \sum_{i=1}^{k} \left[\frac{\left(O_i - E_i \right)^2}{E_i} \right]$$

(2.39).

We then apply it to other probability distributions and the distribution that gives less (χ^2) will be the appropriate probability distribution to present this phenomena or parameter.

Note from the previous equation that when we match distribution resulting from the practical test, the probability distribution be

$$\chi^2 = 0.$$

2.3.4.2   Kolmograv-Smirnov (K-S)

This method is considered to be the second most common method to test how close the distribution is to the probability distribution.

For this method, the distribution is calculated by the assembly output of the test, in cumulative descending order and is in the same table of the previous concrete strength example. The cumulative value is calculated when using the suggested probability distribution and, from the following equation, we can calculate the difference between each of the two values from the suggested probability distribution and from the cumulative test data, according to the number of classes. The biggest difference is calculated in both the number of classes and is often the value of K and S which are selected as the best suitable probability distribution.

$$K - S = \max_{i=1}^{n} \left[\left| \frac{i}{n} - F_X(X^{(i)}) \right| \right] \qquad (2.40)$$

2.4 Decision Tree

The decision tree is one of the basic tools and keys to the managers in each decision-making process, as it is considered a sound and logical way that leads to the selection of the proper decision. Recently, I considered that a person who does not know how to use a decision-making tree is a person living in an isolated cave. Here is proof that this method is the most common way.

The decision tree method is based on the probability of A, B, and C occurring and is calculated by:

$$Ps = P_A \times P_B \times P_C$$

This probability is presented in Figure 2.19. The probability of three events, A, B, and C, occurring at the same time will be presented by the intersection portion from the circle.

The core definition of risk is the probability of the event occurring multiplied by the output of this event: Risk Assessment = probability × consequence.

To explain the decision tree, let's use dice and assume that the first player to throw the dice will earn 6,000 pounds only if the result is a six, but if it is another number, the player will lose the 1,000 pounds and the expected value calculation will be as follows.

The possible emergence of a six is the probability of 1/6, so the expected value equation is as follows:

Expected Value = (1/6) * 6000 – (5/6) * 1000 = 166.7

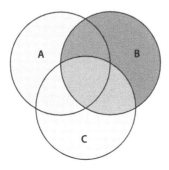

Figure 2.19 Probability theory.

The general equation is as follows:

$$EV = P_s C_s - P_f C_f,$$

where P_s and P_f are the probability of success and probability of failure, respectively. C_s and C_f are the consequence in the case of success and failure, respectively.

We find that the probability of rolling a six is the one-sixth, but the expected value is 166.7, which is greater than zero, which is the expected value of risk. A high expected value indicates that the risk is better. Therefore, when compared with the display of the other player, our player found the expected value greater.

This concept is the main decision-making tool that you would use if you have more than one project and you want to choose one of these projects. Therefore, it is important to be aware of calculating the probability of the success of each project as well as the value of that success.

Therefore, in applying the decision tree method to solve an engineering problem, for example, in the beginning of solving any engineering problem or in the feasibility study, one must specify for the expected outcomes and possible outcomes. To solve the engineering problem at the same time is to determine the likely success of all the possible outcomes. The focus should be to identify all the different ways to determine the likelihood of the event because it depends entirely on the experience from start to finish. Therefore, the experience is the key factor to the success of this method and can be simply explained by the manner of the following example.

This example of Proverbs is common in the case of decisions in engineering projects for the oil industry. You can imagine that all the decisions of drilling for oil depend on the possibility and existence of oil in the ground, and the volume of the amount of ground reservoir varies from a large reservoir to a medium and to a small. Therefore, decision makers should use the decision tree to determine whether the drilling work will do or not.

Figure 2.20 is a case study on making the decision to drill or not. If you drill, you have two possible outcomes – that the well will be dry or that the well will have oil.

If you have an oil reserve, you have three possible outcomes. The reserve may be high, low, or medium. Every outcome has its probability of occurrence. It is noted from the figure that the total probability is equal to one, based on the probability theory. In each scenario of outcomes, calculate the present value (PV).

By multiplying the probability with the present value (PV), one will obtain the expected value (EV) and by adding all the values, you get the

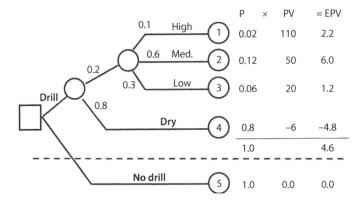

Figure 2.20 Case study one.

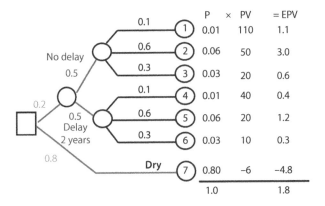

Figure 2.21 Case study two.

expected value of the project which, in this case, is 4.6 million dollars, reflecting the weight of the project, and, therefore, helps you make a decision as to if you will be drilling at this site or direct investments to another reservoir in another location in the world if has the greatest expected value more than 4.6 million.

For example, if you find that your investment in a country gives the expected value of eight million dollars, will you invest in any of the two countries? Naturally, you will not invest in the site in the example, but your decision will be clear that you will invest in the country that will give higher weight to the expected value of invested money. Using the decision tree, potential problems you may encounter during the implementation of a project can be obvious and this is shown by the example in Figure 2.21, which is the same as the previous example with the possible delays in drilling wells. This usually happens in some countries

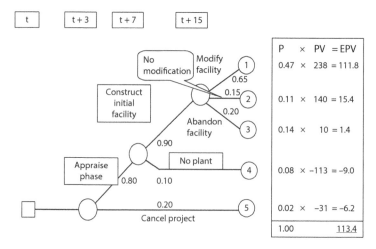

Figure 2.22 Case study three.

that permit administrative bureaucracy where the delay is a result of administrative work for foreign permits and correspondence papers or for the equipment for customs paper work. This is an example where the auger is not available and this is something that will have an impact on the accounts of the present value of investment, as in the following example. It is clear that the number of decision-making trees depends primarily on the experience, as it considers all the possible problems that can be urged as well as prospects.

Figure 2.22 presents another example for constructing a new facility and you may cancel this project or go through the appraise phase (conceptual design). The possible outcomes from the study are that you will need a plant or not. Due to production, there are three possible outcomes: abandon the facility, make no modifications, or modify the facility.

The decision tree is very easy to use, but the problem is how to calculate the probability for every event. You can assume the probability value using your experience, but it will affect the result.

The most proper way to calculate this probability is by using the Monte-Carlo simulation technique. In every case, start by building a model for a reservoir in an oil and gas project, and define the mean, standard deviation, and the probability distribution that are present in the reservoir model, such as area, height, porosity and others parameters. At the same time, build the model for the present value calculation and define the parameters for each variable.

2.5   Monte-Carlo Simulation Technique

Simulation is the process of replicating the real world based on a set of assumptions and conceived models of reality.

The Monte-Carlo simulation is required for problems involving random variables with known (assumed) probability distributions.

This method of simulation was started as an idea by Enrico Fermi in the 1930s. Stanisław Ulam, in 1946, first had the idea and later contacted John von Neumann to work on it and he started to use this simulation in a secret project. After World War II, this simulation was published in many papers as a simulation technique.

The Monte-Carlo simulation technique is frequently used to verify results of analytical methods. Rushedi (1984) used the Monte-Carlo simulation approach to obtain the first two statistical moments (mean, value, and standard deviation) of the failure mode expression of brittle and ductile frames and, consequently, a system safety index. Ayyub and Halder (1985) suggested advanced simulation methods for the estimation of system reliability.

Fellow *et al.* (1993) used the Monte-Carlo simulation program (M-Star) to understand the load and resistance factor design (LRFD). Nikolaos (1995) used the Monte-Carlo simulation to study the reliability of reinforced concrete members strengthened with carbon-fiber-reinforced plastic.

This method depends on simulating the case of study by its parameters and each parameter will be represented by its probabilistic distribution, mean, and standard deviation.

The simulation will have two parameters: a variable and uncertainty. For example, the length of the men in a country is a variable as it represents a normal distribution. But managing a project by time and cost is usually uncertain and is represented by a triangle distribution by knowing the minimum, maximum, and most likely.

So, the risk assessment for the cost estimate and the risk assessment for the project time through the PERT method also uses Monte-Carlo simulation. If you want to predict the cost of a large project, you should break it into parts, define the cost of each part, and add them together. As time management is discussed in Chapter 4, the project time schedule plan is broken into small activities and, based on the PERT method, each activity has a three values as we showed before.

Each random variable is described by its statistical parameters: mean, standard deviation, and type of distribution. The distribution type of the random variable is chosen among the different probability distributions provided by the program.

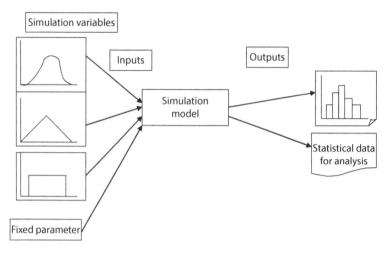

Figure 2.23 Monte Carlo simulation.

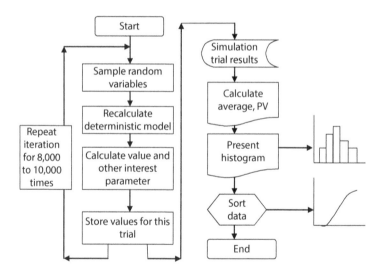

Figure 2.24 Flow chart for Monte-Carlo simulation

Figure 2.23 presents an overview of the Monte-Carlo simulation technique as the input data for the variables will be a probabilistic distribution and, after simulation, will obtain the outputs by the graphs and statistical data.

The simulation model contains all the input data of the deterministic parameters, the random variables, and the equations. The model will run for at least 10,000 trials, as in the following flowchart. The Monte-Carlo simulation technique is simple and is presented in Figure 2.24.

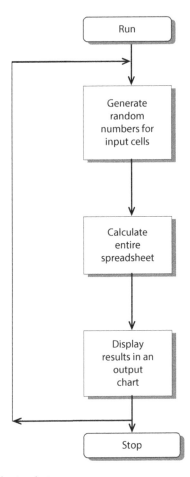

Figure 2.25 Monte-Carlo simulation consequence.

The final output results in displays by software at the end of the simulation and contains the statistical parameters of the variable Z, describing the limit state equation of the cost and time.

After all the trials are completed, this program will calculate and present the mean, standard deviation, and the statistical parameters. Also, it can provide the frequency distribution of the value of the outcomes of Z and determine the probability of increasing the cost to the limit of the budget.

As shown in Figure 2.25, the input data for all the variable parameters is selected by choosing the probability, arithmetic mean, and standard deviation or coefficient of variation. After running the simulation, the output will be a probabilities distribution curve.

Then, run the random numbers as per the "Mid square Method" (Von Neumann and Metropolis, 1940s), which produces pseudo-random four digit numbers:

1. Start with a four digit seed number.
2. Square the seed and extract the center most four digits. (This is your sampling parameter.)
3. Use the sampling parameter as the seed for the next trial. Go to step two.
4. Random number generators usually return a value between zero and one.

For any software you use to perform the simulation for the cost, time, or other risk criteria, the process can be summarized as follows.

Any random variable less than one, in this case this number is assumed to be the cumulative curve value, and by knowing the probability distribution curve, then, by the software definition, the value of that variable corresponds to these random numbers, which correspond to the cumulative value and do that for all variables and put these values in the deterministic model.

From the deterministic model, obtain the value of the output and other parameters. Then, store the data for this trial and repeat these steps again for 8,000 and 10,000 times, so you have 10,000 output for the variable and can calculate the arithmetic mean and standard deviation. Then, draw the distribution or histogram curve and the cumulative curve.

2.6   Risk Adjusted Value (RAV)

Three elements contribute to project ranking: expected return, risk, and loss of funds. One problem concerns trading off each value. Return may indicate one ranking, risk another, and so on. The EPV and risk weighted values, however derived, include each of these components. Compare two projects, each with the same EMV. If the manager could lose ten dollars for a dry hole on one and 100 million dollars on the second, EPV would imply that we would be indifferent between the two. Yet, in the real world, few managers would treat a potential 100 million dollar loss the same as a ten dollar loss. A key assumption in EPV analysis is that the manager risk is neutral and the firm has unlimited capital. Neither assumption holds true in the real world.

Extending EPV to reality requires the concept of certainty equivalence. By definition, certainty equivalent is the value a manager would just be willing to accept in lieu of the risky investment. This point defines the value at which the manager is indifferent between the two alternatives. To illustrate, consider an investment with an EPV of ten million dollars. EPV includes a return, chance of success, and a potential loss. The actual outcome could be higher or lower than this value. Should someone start offering the manager a guaranteed amount less than twelve million dollars, say nine million, then eight million, and so on until the manager said yes, the value eliciting the yes is the certainty equivalent or the indifference value.

Two companies have different, but identical prospects worth ten million dollars (EPV), each with a 100 percent WI, and an offer to farm-out to a third party. The first party accepts a guaranteed offer of six million dollars, while the second party accepts an offer for four million dollars. Why would the indifference points differ? Because most managers are risk adverse, contrary to the primary assumption of EPV and the degree of risk aversion depends on two basic components: the wealth of the firm (hence, freedom from bankruptcy) and the budget level. In this example, both firms are risk averse because they would accept a lower amount to reduce risk. Had either accepted the EPV they would be risk neutral, and occasionally we see investors who would want more (a true risk taker).

Cozzolino (1977) introduced the term risk adjusted value to integrate these concepts, as defined in Equation 2.40.

$$RAV = \left(\frac{-1}{r}\right) Ln\left[P_s e^{-r(R-C)} + (1-P_s)e^{-rC} \right] \qquad (2.40)$$

where r equals risk aversion level of the firm, P_s equals probability of success, R equals NPV of success, C equals NPV of failure, E equals exponential function, and Ln equals natural logarithm.

In the Cozzolino (1977) format examples, like those above that are used to solve for r, assume that RAV is already established. If R, C, RAV, and P are known, the corporate risk aversion can be determined. Without performing an example, larger values for r imply more risk aversion, while smaller values reflect lower risk aversion. Evidence suggests that an inverse relationship exists between capital budget size and risk aversion level. Smaller companies tend to be more risk averse and, thus, tend to spread their risk across as many projects as possible.

This basic format has been extended by Bourdaire et al. (1985) to eliminate the need to estimate the risk component. By employing the elements of subjectivity and assuming an exponential utility function, Equation 2.42 results.

$$RAV = m - \frac{s^2}{2B} \qquad (2.42)$$

where m equals the mean NPV, s^2 =equals the standard deviation of distribution, and B equals total monies budgeted for risky investments.

RAV, under this format, can be based on information typically generated in the evaluation. RAV also depends on the estimated value relative to the dispersion of the NPV outcome. More importantly, high dispersion projects may be ranked above projects with lower standard deviation if the dispersion relative to the budget is low. RAV depends on two basic relationships: m relative to s^2 and s^2 relative to B.

If the mean value of NPV that was calculated from the Monte-Carlo simulation is equal to ten million dollars, then a standard deviation illustrates the dominance of the dispersion term, since $10 - (15)^2$ will be a very negative value. Now, suppose that the investor is a large oil company with a budget of $1,000 million. The RAV of the project is

$$RAV = 10 - \frac{225}{2 \times 1000} = \$9.9 million.$$

For a smaller investor with a budget of only $200 million, RAV becomes

$$RAV = 10 - \frac{225}{2 \times 200} = \$9.4 million.$$

The breakeven RAV value for B is found by solving

$$RAV(breakeven) = \frac{s^2}{2 \times m} = \frac{225}{2 \times 10} = 11.25.$$

We are not aware of anyone presently ranking on RAV, although more people are discussing it. Like other ideas portrayed in this book, we believe it should be included as part of the evaluation for a period of time. If RAV aids in decision-making, then include it permanently.

3

Pitfalls in Time Schedule Planning

3.1 Introduction

The initial and basic principle of project management is how to read the time schedule, which is a real representation of the project performance with the expectation of what could occur as a result of implementation.

The first important step is to determine the purpose of the project, which must be carefully defined, and then answer the following question: what is the project driving force – time or cost?

For over 20 years, schedules have been prepared manually. Now there are computer programs that can be used to deliver time schedules.

In general, there is more than one method for drawing up a time schedule for the project and this happens according to the nature of the project and the required presentation that will be provided to senior management.

We must recognize that the preparation of the schedule is the cornerstone in the management of projects and will follow the work schedule of the allocation of human resources and equipment, as well as the distribution of costs along the project time period with the identification of ways to control the costs.

In the 1900's, during World War I, Henry L. Gantt used the first method to prepare a project schedule. This is considered the first scientific method for the preparation of schedules.

This method is simple in the representation of activities by rectangles is used in project planning and work schedules at the time of production. The Gantt Chart was used by putting a plan on a magnetic blackboard using rectangles of iron, whose lengths were time units.

This was developed to be the S curve and is considered the first method to follow up the project with different activities by distributing the resources on the activity which one can use to monitor the performance.

Until the mid-fifties, there had been no mention of any development in project planning. In 1957, there were two different teams working on project planning using networks.

The first team was prepared by way of Program Evaluation and Review Technique (PERT). This method depends on the probability theory.

The second team was using a network and depended on CPM (Critical Path Method), and methods of those networks have the same methodology but a difference in objectives. The development is done through operations research.

The first team started using the PERT method when the United States Navy was faced with the challenge in the POLARIS system when they wanted to make rocket launchers in record time in 1958.

The basics of the PERT method are to overcome the lack of defining the activity duration time exactly and to use statistical methods to calculate it. This is done by defining the maximum, minimum, and most likely time for each activity in order to finally obtain the likelihood of the completion of the project or important parts of the project and the minimum and maximum probable time to finish the project.

The teamwork on the CPM was introduced in 1957 by two companies: Du Pont and Remington Rand Univac.

The objective of the working group is to figure out how to reduce the time period for maintenance and overhaul the rotating machines, as well as the construction work.

It is noted that the calculation of the time required for different activities can be in CPM more easily than the POLARIS project activities, as we need to identify one expected time period only for each activity and the longer timetable for the course of the series of activities has been defined as the critical path.

Now, the critical path method is the most common way of networking activities in project planning. It is used with some other methods by utilizing computer software.

The process of planning is simply to plan what will be done in the project in accordance with the order and manner in the execution of the project. There will often be some changes, and you have to adjust the time schedule in accordance with the changes in the project.

In order to do the work with good planning, you must answer the following questions clearly:

- What are the activities that you want to execute?
- When will you execute these activities?
- Who will execute these activities?
- What are the equipment and tools required?
- What activities cannot be executed?

The answers to the previous questions are the key to arranging the work in an appropriate way. From that point, the project will be understandable. Now, your goal is to transform this information in a simple way, present it to all parties of the project, and make sure everything is clearly understood.

Your planning team target is to implement the project in a timely manner in accordance with the specific cost and, at the same time, achieve the required level of quality. Therefore, the planning of this project is needed for the following:

- To reduce the risks of the project to the lowest level possible
- To achieve the performance specifications of the project
- To establish organization for the implementation of business
- To develop procedures to control the project
- To achieve the best results in the shortest possible time.

The planner cannot plan, in detail, every minute of the project due to the non-availability of all information, but, with time and the execution of work, some of the details will have to increase the effort and increase the time for action to adjust the timetable in accordance with the new information and details that he or she will obtain.

If you ask people what makes a project successful, "a realistic schedule" usually tops the list. But, ask them to be more specific and several characteristics of a realistic schedule emerge. A realistic schedule does the following:

- Includes a detailed knowledge of the work to be done
- Has task sequences in the correct order
- Accounts for external constraints beyond the control of the team
- Can be accomplished on time, given the availability of skilled people and enough equipment.

Finally, a realistic schedule takes into consideration all the objectives of the project. For example, a schedule may be just right for the project team, but if it misses the customer completion date by a mile, then it's clear the whole project will need reassessment. Building a project plan that includes all the necessary parts and achieves a realistic balance between cost, scheduling, and quality requires a careful, step-by-step process.

3.1.1   Plan Single Point of Accountability (SPA)

This will be done by the project manager with the planning team. First, define the team members that will perform the required activities and be sure they have sufficient information on their potential and their relation to the size of the project. If you want to use another experienced planner from another project contracted to work with you, do so at the beginning of the project.

You should also know through the collection of information whether the working group has worked in similar projects because projects have almost the same activities. For example, if you are working on an oil and gas project and the working group has worked on the same type of project before, that experience is different than housing projects, hotels, road projects, or administration buildings.

Each type of project has its own characteristics. Therefore, the working group needs to have worked in a project that is similar to your project.

The planner must be efficient in planning, must have the ability to plan the project well, and must have good experience in the same type of project.

At the beginning of the work, it is very important to hold a meeting between the planning team, the official sponsor of the project or the director of the project, and the owner and his or her representative. The goal of this meeting is to clarify the main objectives of the project, identify priorities in the implementation of the driving force, which is time or cost, and determine what is desired from the project as a whole.

3.1.2 Starting the Plan

Before starting the project plan, we should go through the basic definitions that are usually used during plan implementation. These definitions are as follows:

- Task – a small amount of work that will be performed by one member
- Activity – consists of a set of tasks and is performed by different individuals
- Concurrent activities – activities that are performed in parallel
- Series activities – activities that are executed one after the other, as the second activity cannot start until the first activity is finished

Usually, there is a conflict between the task and activity and it can be clear if you have to prepare a technical report. An activity consists of tasks, such as collecting required information, performing a data analysis, preparing photos and figures, preparing the first revision of a report, or printing a report.

There are many ways to start planning and you have to choose among them. A good way to begin is by identifying the key stages of the project. The main key stages of the project can be identified by holding a meeting with the experienced people on the project team from different disciplines, stakeholders, and sponsors. In this meeting, use a brainstorm technique among the attendees.

Every group should make suggestions on paper. Then, the papers should be collected and all the ideas and contributions, regardless of being logical or illogical, should be shared among everyone in the meeting. It is important to follow the following rules during the meeting:

- Be concerned about quantity and not quality, even if it turned out that some of the tasks and activities have been replicated.
- Stop any suspicion of an individual to avoid any idea of the critical observations bothering the participants.

The next step is very important, as it will now have a wide range of tasks. The next step of the action team is to filter such activities and this is done by removing some of the tasks that are repeated or duplicated. Compile the tasks, including the interdependence of both the straight or parallel. The small number of tasks and activities reduced often ranges from 30 to 60, according to the size of the activity of the project. Then, compile the activities at key stages of the project.

By using this method, you will reach high precision in the planning, up to about 90 percent. This is considered the beginning step in the planning of the project as a whole.

Now, you have the main stages of the project and all the key stages were agreed on by the members of the project. Now you should order them in a logical order, but you should avoid the following:

- Defining time or dates.
- The allocation of employment to those stages.

All of the above will cause problems, as many attendees will push you to define dates. Please take care to not fall in this hole.

To avoid mistakes in planning the Project Logic Control, the key stages must be defined on the main wall of the office. Figures 3.1a and 3.1b show examples of the main stages of the project.

The advantages of the above method are that everyone has an opinion on the project, making everyone keen to the success of the project, and a person's idea or opinion reflects on the project. Therefore, a person will do his or her best to offer opinions that will match with the project goal.

It is noted from Figures 3.1a and 3.1b that the design phase has been divided into two stages, the first stage being (a) and the second being (b), in order to allow the sending of purchase orders from the start before the end of the first phase of the design.

Now you have the information that can be used in the computer software in order to prepare the time schedule for the agreed plan.

The basic rules that must be adhered to and strictly followed in the preparation of a project schedule are as follows:

- The movement of activities should go from the left to the right.
- There is no measure of time.
- There is a place to start in the beginning of the greatest square in the north. Make sure there is an empty place in the page for each major stage in the project.
- Each phase is described by the act of writing in the form of present tense. (Do not try to set the stage for any period of time.)
- The pages are developed in accordance with the logical arrangement.

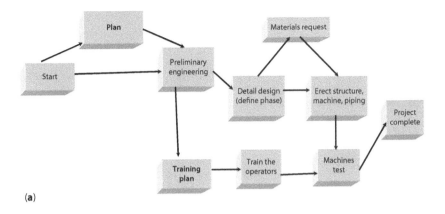

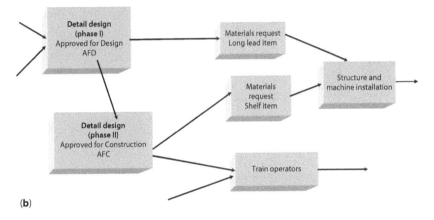

Figure 3.1 Project key stages.

- There must be communication between the stages of a relationship.
- Identify responsibilities.
- Provide connectivity between the stages.
- Avoid the intersection of the stock as much as possible.
- Identify each key stage by professional codes.

3.1.3   Work Breakdown Structure (WBS)

The work breakdown structure (WBS) is the most important issue in the project plan. The WBS defines the work that has to be done to complete the project. Moreover, the WBS can help determine the cost of the project and its schedule.

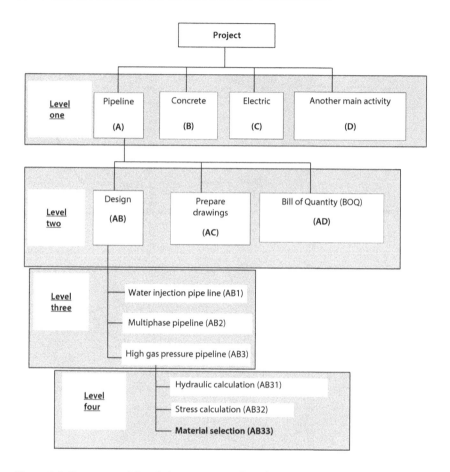

Figure 3.2 Presents work break down structure (WBS).

As commonly used in projects, as in Figure 3.2, a project consists of three pipelines: a water pipeline to transport water, a pipeline for crude oil production from the plant to storage, and a gas pipeline to transport gas from the production plant to the treatment plant.

The structure of the work activities are divided by the project to more than one level. In the first level, define the main stages of the project and then in level two and level three, as in the previous example of the pipeline, there are three pipelines intended for their establishment. Concrete work is the concrete base, as well as electrical work, which is the controller of the remote and can thus write the main stages.

The second level in the WBS will be focused on in the Figure 3.2 on the stage in the pipeline work where there will be more than one stage for the design drawings, calculations, and materials bill of quantity.

Level three, in this example, focuses on the design phase and will be divided for the design of the gas pipeline, oil, and water.

In the fourth level in the example, we design a gas pipeline by performing hydraulic design and pipe stress analysis to choose the thickness of the pipe and support locations and to ensure that the subsequent stage is the selection of the required valves. In some projects, it may require several stages depending on the nature of each project.

We note each stage in each level code for ease of use later. The WBS can be completed at any level of description. The WBS does not explain the relations between the activities nor does it show the time or the time period for any activity.

In summary, the WBS is implemented by following the steps as shown in Table 3.1.

After you have selected the main stages and WBS, the next step is to develop a rough period of time on the schedule, but there is an important step prior to that, which is to define responsibilities.

21-Your organization is having a difficult problem in time management for all of its projects. The CEO asks you to help senior management get a better understanding of the problems.

What is the FIRST thing you should do?

- Meet with individual project managers to get a better sense of what is happening.

Table 3.1 Work breakdown structure.

What	The WBS is a high-level breakdown of work scope, a list of main project deliverables, and can be broken down by materials, contracts, area, or defined work packages.
Why	The WBS is used to break jobs into linked tasks. It is the basis for the estimate, the cost report, and the execution plan. Having a common format across all elements of the project results in simpler cost tracking and forecasting.
How	The project team should brainstorm the best way to control and implement the project by assessing the project execution methodology together with the commissioning sequence.
When	The WBS should be included in the project execution plan, early in the select phase, and revised throughout to define and execute.
Who	Project Leader, Planner, SPA, Construction Engineer, Commissioning Engineer, Estimator

- Send a formal memo to all project managers requesting their project plans.
- Meet with senior managers to help them develop a new tracking system for managing projects.
- Review the project charters and Gantt charts for all projects.

3.2   Responsibilities of the Team

In general, for industrial projects and, specifically, petrochemical projects, there is usually a modification, which will be a minor project, such as constructing a new tank, new compressor, pumps, or other mechanical equipment. The team is usually formulated from different disciplines from different project departments for this one project.

The planning team, who is responsible for preparing the schedule at this stage, has a vital role to distribute the main stages of the project to members of the team. Every main stage has a key stage owner (KSO), whose responsibility is to achieve the required targets within a reasonable time. The responsibilities of the KSO include the following:

- Identify the work to the level of small tasks
- Identify relations between activities and tasks and clearly define them
- Estimate time with a high accuracy
- Ensure that business is done in a timely manner in accordance with the required quality
- Ensure that work is proceeding in accordance with the procedures and requirements for quality assurance
- Maintain ongoing follow-up
- Compose periodic, accurate reports.

As the project or construction manager, you will face problems during a project and you should resolve them in an accurate time. These problems will be mainly as follows:

- The necessary authority to complete the work
- The necessary tools to complete the work
- The right atmosphere to achieve the required quality of work
- Direct support from the project manager or the official sponsor of the project
- The performance expectation is clearly understood.

When choosing the KSO, the following should be considered in the potential KSO:

- Skills
- Depth of information and knowledge
- Previous experience in the same area
- The time which is required to complete the work
- Accuracy in the completion of previous work
- Ability to solve problems
- Ability to manage time
- Ability to work individually and with a team
- The volume of work and the current work of the KSO
- Ability to take and give advice and support
- Ability to work under stressful conditions
- The necessary training required now and in the future

Now the key stages plan is available and every key stage owner has defined responsibilities. Therefore, it is time to start estimating the time needed for each activity in the key stages.

3.3   Expected Activity Time Period

To define the required time to finish any task, you must know the resources available to perform the work according to the required quality and to do that you must know the following:

- Know the task volume and if it can be measured. It will be better, for example, to calculate the time required to prepare the wood form to pour 100 m³ of concrete into the foundation. From that we can figure the task volume.
- Define the work required by hours, days, or weeks to finish this activity, noting that the number of workers should be identified, and consider the capability for each worker to perform the task alone.

Measuring the working capacity for each worker will usually take days. Take care from traps, such as the idea that you should decrease the capacity of work per day by about 50 percent, as all the hours per day do not focus on the project's activities as there is a lot of time wasted in meetings, special discussions, restroom breaks, eating, and others. Moreover, there is some delay in the work itself.

Now define the time period for each task, but take care from other traps when putting the schedule in the calendar because the total time period will be different due to the following factors:

- Weekends
- Official vacations and holidays
- Annual leave for employees
- Some days the project will stop.

It is worth mentioning that defining the performance rate of each activity depends on a normal rate, which is found in textbooks or standard guidelines for some contractor companies. But it is essential to take any information from others who have extensive experience, work in the same country or the same location, or have experience working on similar, previous projects.

If the same activity in similar projects, such as pouring concrete foundations, was repeated before, but we must choose a foundation of the same kind, it is preferred to be from the same type and the same location because, for example, remote desert areas are different than cities where there is labor available and the efficiency of employment is usually higher than the remote area. For industrial projects, most of the projects are outside of the big cities, in the desert, or in remote areas and all the contractors and engineers main offices are from the city where housing and normal building are famous activities. The planner should take care from the difference of the project location.

When you need information from individuals or experts, remember that not everyone has accurate information, as by the human nature. When someone says that the expected time for the activity was eighteen days, but he finished it in ten days, only trust this happened because he considered it one of his success stories, so it will keep in his mind for a long time.

As a rule of thumb, no one can work 100 percent of his or her time because about 20 to 50 percent of time is wasted in the following activities:

- Attending meetings that we don't really need to attend
- Spontaneous office visits
- Opening and reading mail and email
- Searching for specific information
- Providing assistance or advice to others
- Equipment failure such as a computer, printers, and others
- Daily regular activity

- Misunderstanding between team members
- A lack of clear specification or scope of work
- New specification to quality
- Attending a training course or seminar

3.4   Calculate the Activity Time Period

The calculation of the activity period requires taking data from different disciplines, but its accuracy depends on the skill and strong experience of the planner. On the other hand, there is a performance rate for each activity and, in international contractor companies, usually have a standard guide for the performance rate for each activity.

There are many factors that affect labor productivity:

- Job size and complexity
- Job site accessibility
- Labor availability
- Equipment utilization
- Contractual agreements
- Local climate
- Local cultural characteristics, particularly in foreign operations.

There is a method for calculating the time of an activity through the use of performance rates. Below shows the method of calculating the period of time to excavate 15000 m³, and this requires, first, determining the method of excavation. In this example, it has been identified by the equipment.

Assume you need to excavate 15000 m³ of soil. Using two bulldozers and loaders and trucks to transport the disposal will be the method of excavation. The equipment used will be two bulldozers, two loaders, and four trucks. The performance rate for one bulldozer is 120 m³/hour and the performance rate for one loader and two trucks is 75m³/hour.

The rate of excavation is $2 \times 120 = 240$ m³/hour.

The required time for excavation is $15000/240 = 62.5$ hours $= 9$ days.

The rate for removing disposal is $2 \times 75 = 150$ m³/hour.

The time period for removing disposal is $15000/150 = 100$ hours $= 15$ days.

Therefore, the excavation time equals about nine days and the transfer of residue is approximately fifteen days, taking into consideration that the number of working hours in the day equals only seven hours in the preparation of the timetable. For the excavation activity, take the largest time period which is fifteen days. It may be after finishing the excavation to start pouring the plain concrete, so it will start after nine days, as there is no need to wait pouring the concrete until the transfer of all waste from the site is completed.

3.5   Time Schedule Preparation

Now you have a plan for the work and the time period for each activity. The most important thing in preparing the time schedule is to identify the activity and determine the time required for its implementation with the identification of relations between the activities. How to determine the activities and knowledge of time of each activity has been presented, so we must now know how to arrange the activities.

The different common relation between activities is as shown in Figure 3.3.

1. Activity (B) cannot start until the finish of activity (A).
2. Activity (A) and (B) start at the same time.
3. Activity (A) and (B) finish at the same time.

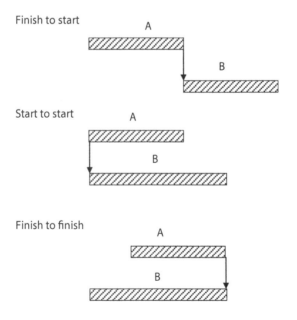

Figure 3.3 Activities order.

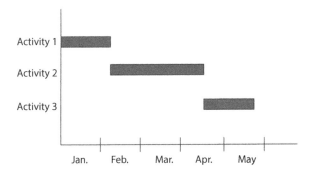

Figure 3.4 Example for a Gantt chart.

In some cases, activity A may begin and, after a certain time period, activity B begins. This period is called the "lag time."

A computer can perform this task easily, but, in general, there are two main methods: the arrow diagram and the precedence diagram, as shown in Figure 3.4.

3.5.1   Gantt Chart

In general, the relations between activities are usually in series or parallel. The relation depends on the nature of the activities and this should be considered in preparing the time schedule.

The Gantt chart is the oldest and most traditional method used to present project schedules to this day. But, the relations between all the activities are not presented well by this method and you cannot go through the detailed activity as accurately. Therefore, it would only be useful in presentations for high-level management, as it presents enough information for level one of the schedule, but for more detail another method is needed.

3.5.2   Arrow Diagram Method (ADM)

This diagram depends on the definition of each activity by an arrow, as shown in Figure 3.4, and the point of connecting arrows called nodes, which are drawn as circles. This method is also called Activity on Arrow (AOA).

In Figure 3.5, activity A depends on activity B. However, activity C depends on B and D, but there is no activity line between D and C. So, put in a dummy arrow (with dashed line) with time zero to solve this situation. Dummy activities for large activities can present a problem, so the use of a precedence diagram is preferred.

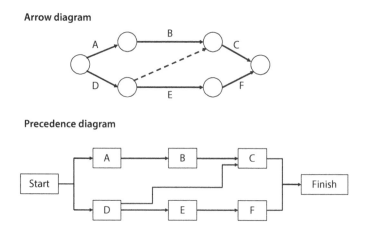

Figure 3.5 Tools to arrange the activities.

3.5.3   Precedence Diagram Method (PDM)

This method is the most common one. In this method, every activity is presented by a box or a rectangle, as shown in Figure 3.5, and the detail of the rectangle is shown in Figure 3.6. The rectangles are connected by the arrows which represent the dependencies between the activities. From this figure, it is clear that activity C starts after activities D and B.

Figure 3.6 presents the inside of a rectangle, which will note the duration, early start time (ES), early finish (EF) time, and also the latest start (LS) and latest finish (LF) time.

3.5.4   Critical Path Method (CPM)

The essential technique for using CPM is to establish a model of the project that includes a list of all activities required to complete the project, the time duration that each activity will take to completion, and the dependencies between the activities.

By using these values, CPM calculates the longest path of planned activities to the end of the project and the earliest and latest time that each activity can start and finish without making the project take a longer time.

This process determines which activities are "critical," which are the activities on the longest path, and which have "total float," as these activities can be delayed without making the project longer.

Any delay of an activity on the critical path directly impacts the planned project completion date (i.e. there is no float on the critical path). A project can have several critical paths. An additional parallel path through the

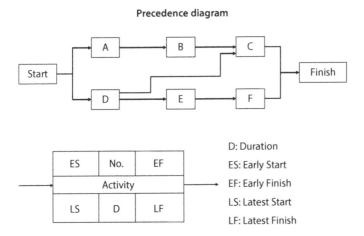

Figure 3.6 Method of preceding diagram.

network with the total durations shorter than the critical path is called a sub-critical or non-critical path.

3.5.5 Program Evaluation and Review Technique (PERT)

This is required in any case of special projects or large-scale deliveries with a variety of activities. All activities in the project need to be identified accurately and all must be completed before the project completion. These activities must be initiated and carried out in a specific sequence and must be completed before the start of the completion of other activities. Some activities may occur in parallel, meaning that two or more could be completed at the same time.

In addition, there are specific achievements that indicate the completion of key stages in the project. The successful management of such projects must be carefully planned during implementation of these projects to speed up the delivery during a specified period of time, to coordinate the multi-activity of the project and monitor the use of various resources necessary for its implementation, and to achieve the project on time within the budget cost.

One of the operations research methods, which are used to assist the planning team, the coordination, the control of such projects special in case of the multiplicity and complexity of activities, and large size projects is known as the method of Program Evaluation and Review Technique known as PERT.

Although the method of PERT is mainly used for military purposes, it has been used successfully since 1959 for most large-scale projects. PERT analysis is used in a number of areas of the computer industry, construction industries, and in planning the shutdown for maintenance in refineries.

This analysis has confirmed its applicability and importance by its application in different projects. Contracting companies can use it successfully because of its role in solving problems of coordination between various activities in a project that have considerable degrees of complexity and its role in planning the time required for implementing each activity. This method has made it possible to complete a project within the time planned and the overall schedule.

PERT was developed primarily to simplify the planning and scheduling of large and complex projects. It was able to incorporate uncertainty by making it possible to schedule a project while not knowing precisely the details and durations of all the activities.

Hence, the PERT method depends on using statistics and probability theory, so it is now the main key for the project risk assessment from the time schedule point of view.

In some cases, it is required to compress the time schedule and items by a crashing technique as to reduce the time and calculate its impact out the cost and let them work overtime. The other way is by using a fast tracking method. This is done by letting some activities be in parallel which are usually in series. As we discussed, in the oil and gas industry, time is very critical. For example, we can start to construct the foundation for a petroleum processing plant before finishing 25 percent of the engineering deliverables. Fast tracking often results in rework and usually increases risk.

Any dependencies between activities require specification of a lead or lag to accurately define the relationship. An example of lead in a start-to-finish dependency with a ten day lead is in a case where the successor activity starts ten days before the predecessor has completed. An example of a lag is when there might be a desire to schedule a two week delay (lag) between ordering the equipment and using or installing it.

3.5.6   Example

The following example will illustrate the relationships between activities and how can you create them through the precedence diagram.

The example in the following table shows the activities for a cast concrete foundation under machine package and connects the machine with piping to the facilities, which is a simple example of knowing how to arrange activities, account for the overall time of the project, and identify the critical path.

Table 3.2 Example for foundation.

Item	Activity	Time (days)	Precedence Activity
100	Mobilization	3	–
200	Excavation	8	100
300	Pouring concrete foundation and piping support	10	200,100
400	Install the piping	4	300,100
500	Install the mechanical package	1	300
600	Put the grouting	1	300, 500
700	Connect the piping	5	400,500
800	Commissioning and start-up	2	700

In the above table, the first column contains the item number or the code for that item according to the project and its activities and the sub activities. Also, the main activity code may be 100 and the sub activity may be 110 and so on, but the above example is a simple case.

The second column is the name of the activity, which describes the activity, and the third column is the time period for each activity in days.

The fourth column specifies the relationships between activities. As for activity No. 200, which is the excavation, it is constrained by activity No. 1. Activity No. 6, which is setting the wooden form, is related to two activities, No. 300 and 500, meaning that the setting of the wooden form depends on finishing the preparation of the wooden form (item No. 300) and pouring the plain concrete (item No. 500).

Figure 3.7 shows the precedence diagram and in each diagram there is a number with its time duration. Figure 3.8 shows the early start and finish for each activity by using the following equation:

$$EF = ES + D,$$

where EF is early finish, ES is early start, and D is duration.

Start with activity number 100. The early finish (EF) of this activity is three days. Transfer this value to activities 200 and 400 in the early start (ES) rectangle zone.

Item number 800 depends on 600 and 700, so it will take the higher value in the early finish (EF), which is 23 days. Put this number as the early start (ES) for activity 800.

For the last activity, which has an early finish at 25 days, transfer this value to the latest finish (LF) rectangular zone.

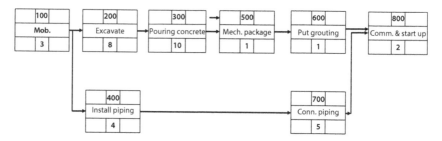

Figure 3.7 Precedence diagram for the example.

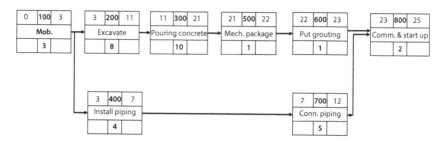

Figure 3.8 Calculate the early times.

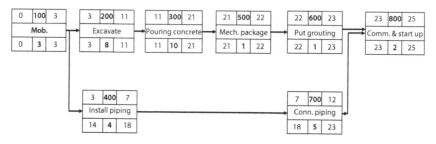

Figure 3.9 Calculating the latest times.

Figure 3.9 shows the latest start and finish for each activity by applying the following equation:

$$LS = LF - D.$$

In the last activity, calculate back because the latest early finish will be 25 days by applying the above equation.

From Figure 3.9, by subtracting the latest start (*LS*) from latest finish (*LF*) and subtracting the early start (*ES*) from early finish (*EF*), it gives the value of zero, so it means that this activity is on the critical path. But, if the difference of the subtraction has a value, it means that this activity can be delayed by this time period without affecting the total project time.

Table 3.3 Float time calculation.

No.	Activity	D	Earliest		Latest		Float		Critical
			ES	EF	LS	LF	TF	FF	
100	Mobilization	3	0	3	0	3	0	0	*
200	Excavation	8	3	11	3	11	0	0	*
300	Pouring concrete foundation and piping support	10	11	21	11	21	0	0	*
400	Install piping	4	3	7	14	18	11	0	
500	Install the mechanical package	1	21	22	21	22	0	0	*
600	Put the grouting	1	22	23	22	23	0	0	*
700	Connect the piping	5	7	12	18	23	11	11	
800	Commissioning and start up	2	23	25	23	25	0	0	*

When we calculated the early and late schedule dates for our project, we found that sometimes the early and late schedule dates were the same and in other activities the dates were different. In these activities there was a difference between the earliest day that we could start an activity and the latest day we could start the activity. The difference between these two dates is called "float," or sometimes "slack." These terms mean exactly the same thing and can be used interchangeably. The float of an activity is the amount of time that the activity can be delayed without causing a delay in the project.

Using computerized project management scheduling software, we can modify the list of activities on the critical path to include activities that are nearly on the critical path. This is important since the critical path method is a method for managing project schedules. The activities that have zero float are the activities that cannot be delayed without delaying the completion of the project. These are the activities that must be monitored closely if we want our project to finish on time. Conversely, the activities that are not on the critical path, those activities that have something other than zero float, need not be managed quite as closely. In addition, it is important to know which activities in the project may be delayed without delaying the project completion.

Resources from activities having float could be made available to do a "workaround" if the need should arise.

By performing a simple calculation, we can find eleven days as a total float (TF) for installing piping. But, installing piping has zero free float (FF), as any delay will affect the connection of the piping. The piping connecting activity has an eleven day FF, as there is no activity after that to delay.

3.5.7 Application of the PERT Method

We previously reported that the implementation of the activity takes time and needs resources. The PERT team may face a problem of how to estimate the time of each activity, as they see that reliance on a single estimate time is a weak assumption and is incompatible with the conditions of uncertainty, even under the best circumstances. The external environmental factors surrounding the project may cause a deviation in the times of the activities planned.

Accordingly, the PERT method has been found to overcome the time uncertainty. By estimating the time of each activity with three estimated time values they will be gathered together with knowing the activity estimate and time probability distribution curve to reach a statistical probability to estimate the time of the project's completion. This method will apply as follows.

Optimistic time estimates the minimum possible time in which to do the activity, considering that all the factors affecting this activity are going smoothly. This time is usually low because all the circumstances must be good at the same time.

Pessimistic time estimates the maximum possible time in which to implement the activity. It considers the worst cases that happen to delay this activity. This usually provides a higher time period with low probability.

The "most likely time" presents the time period in which to perform this activity under normal conditions for all the factors affecting this activity. This value has the most probability of occurring.

The PERT method correlates the relation between the above three times by using a beta distribution as in Figure 3.10. The optimistic and pessimistic value precincts the limits of the distribution, but the most likely value present is the frequencies of the event occurring.

To define the activity time period, the activity mean time needs to be calculated, and this is done using the below equation. The summation of all the time periods on the critical path presents the mean time period for the project.

3.5.7.1 Statistics Calculation for Activity Time

From the beta distribution, the expected or average time for each activity is calculated according to the following equation:

$$T_a = \frac{T_o + 4T_m + T_p}{6},$$ (3.1)

where T_a is average time, T_o is optimistic time, T_p is pessimistic time, and T_m is most likely time.

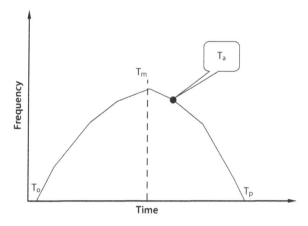

Figure 3.10 Probability distribution for every distribution.

Table 3.4 PERT example.

To	Tm	Tp	Activity	Item
1	3	5	Mobilization	1
6	8	11	Excavation	2
8	16	25	(Pouring concrete foundation and piping support)	3
3	4	5	Install the piping	4
1	1	2	Install the mechanical package	5
1	1	3	Put the grouting	6
2	5	7	Connect the piping	7
1	1	2	Commissioning and start up	8

3.5.7.2   *Example*

For the previous example of constructing the concrete foundation, there are three values for time for each activity, as shown in the following table.

After defining the critical path from the previous example, calculate the expected time at each activity as shown in Equation 3.1.

Calculate the standard deviation using the following equation:

$$s = \frac{T_P - T_o}{6} \tag{3.2}$$

Variance, V, is S^2.

From Table 3.5 calculate the Ta, which is the average time and the standard deviation based on Equations 3.1 and 3.2, respectively. The variance

Table 3.5 PERT calculation.

V	S	C.P.	Ta	To	Tm	Tp	Activity	Item
0.44	0.67	*	3	1	3	5	Mobilization	100
0.694	0.83	*	8.3	6	8	11	Excavation	200
1.36	0.5	*	7.17	8	10	15	(Pouring concrete foundation and piping support)	300
0.11	0.33		4	3	4	5	Install the piping	400
0.028	0.167	*	1.17	1	1	2	Install the mechanical package	500
0.11	0.333	*	2.17	1	1	3	Put the grouting	600
0.694	0.833		3	2	5	7	Connect the piping	700
0.111	0.333	*	1.5	1	2	3	Commissioning and start up	800

is calculated as it is only possible to sum the variance. You cannot perform summation for the standard deviation.

The minimum time to finish the project is eighteen days and the maximum time to finish is 39 days. The average time period to finish the project is in 23.31 days.

What is the probability if you increase the project time over 25 days? This value can be calculated from the following equation:

$$V = 0.44 + 0.694 + 1.36 + 0.028 + 0.11 + 0.11 = 2.74 \text{ days.}$$

Standard deviation (S) is 1.66 days.

$$z = \frac{26 - 23.31}{1.66} = 1.62$$

From the probability distribution tables, the probability that the project completion period is more than 26 days is five percent. The probability that the execution time for the project is equal to or less than 26 days is 95 percent, as shown in Figure 3.11.

3.5.7.3   Time schedule control

For real projects, four legs platforms in a water depth around 80 m, the percentage of hours of different disciplines will be as follows in Table (3.6) as a guideline:

The approximate man hours for structure engineering activities is about 5600–6300 man hours, whereas the total project man hour is about 38000

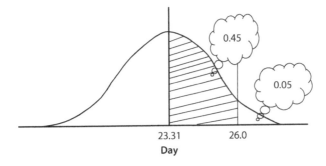

Figure 3.11 Probability distribution.

Table 3.6 Percentage of man hour in engineering phase

Discipline activity	Percentage of the man hour
Project management	25%
process	12%
Mechanical	7%
Piping and layout	16
Electrical	8%
Control and Instrument	3%
Structural engineering	17%
Pipeline	5%
HSSE	7%

to 44000 man hours. Noting that man hours reflects to money, the cost is around, for this project, two million USD. As if you have a high fee rate per man-hour, the man-hour may be reduced to less than above. It depends on how the engineering firm approaches their proposal as some engineering firms reduce the number of man-hours and increase the cost per man-hour and vice versa according to their strategy and the market share in this time period.

The progress of an engineering activity is required to be identified for control, so you should put some basics to control. If the engineering firm will do the work, the progress measurement should be identified and be agreed upon by the client and the engineering firm before starting the project. For example, in a case of study reports, the following is an example for the progress measured as shown in Table 3.7.

For different engineering discipline activities, the measured progress should be defined. As in oil and gas projects, preparation of P& ID is very

Table 3.7 Engineering report progress measurement guideline.

Activity	Progress percentage
Issue draft to the client	30%
Completed document review by the client	50%
Incorporate the client comments	80%
Final report approval	100%

Table 3.8 Engineering preparation for P&ID.

Activity	Progress percentage
Study and prepare draft or sketches	20%
Completed document review by the client	60%
Issue for design	75%
Issued for construction finally without hold	100%

Table 3.9 Procurement progress measuring.

Activity	Progress Percentage
Issue RFQ to bidders	20%
Bid evaluation and selection	30%
Issue purchase order to vendor	40%
Approval of vendor prints	50%
Materials successfully passes tests/inspection	85%
Materials ready for FOB	90%
Materials received on site	100%

critical, so the most practical progress measurement will be as follows in Table 3.8.

For procurement services, the following Table 3.9 presents the traditional progress for measuring percentages.

The requisition of bulk materials like steel structure, piping, electrical cables, or others is based on the material take offs (MTO) and, in most cases, it will be issued in three MTOs. The following weights are to be applied for bulk MTO:

First MTO requisition 65%

Second MTO requisition 25%

Third and final MTO requisitions 10%

For the equipment delivery, mile stone progress will use as Table 3.10 as a guideline. It defines the percentage progress after finishing the mile stones.

Table 3.10 Manufacturing delivery progress measuring guideline.

Manufacturing and delivery milestone	Equipment	Bulk materials
All vendor drawings and vendor data approved	10%	
All sub- ordered materials received by vendor	25%	10%
Materials successfully inspected, tested and accepted	50%	75%
All materials delivered on site	15%	15%

Table 3.11 Overall project progress measuring guideline.

Activity	Percentage
Detailed engineering	10%
Procurement services	5%
Delivery of equipment materials	25%
Construction	25%
Installation	25%
Professional services	10%

For overall project progress, the following measurements in Table 3.11 can be considered as a guide line.

3.6   Planning Overview

The following steps are important to obtaining the schedule plan:

- *Create the project definition.* The project manager and the project team develop the statement of requirements (SOR), which identifies the purpose, scope, and deliverables for the project and defines the responsibilities of the project team.
- *Develop a risk management strategy.* The project team evaluates the likely obstacles and constrains and creates a strategy for achieving the required costs, schedule, and quality.
- *Build a work breakdown structure.* The team identifies all the tasks required to build the specified deliverables. The scope statement and project purpose help to define the boundaries of the project.
- *Identify task relationships.* The detailed tasks, known as work packages, are placed in the proper sequences.

- *Estimate work packages.* Each of these detailed tasks is assigned an estimated for the amount of labor and equipment needed and for the duration of the task, which will be explained in Chapter 5.
- *Calculate the initial schedule.* After estimating the duration of each work package and figuring in the sequence of tasks, the team calculates the total duration of the project. This initial schedule, while useful for planning, will probably need to be revised further down the line.
- *Assign and level resources.* The team adjusts the schedule to account for resource constraints. Tasks are rescheduled in order to optimize the use of people and equipment used on the project, which will be discussed in Chapter 5.

These steps provide all the required information to understand how a project will be executed. The steps are systematic, but they don't necessarily come up with the "right answer." It may take several iterations of these steps to find this answer, which is the optimal balance between cost, schedule, and quality.

The planner plays a key role in controlling the project outcome and flagging potential bottlenecks and problems for the project manager. It is expected that the planner set up a weekly review meeting for each of his or her projects with the following deliverables:

- Attendees should consist of the appropriate cost engineer, project manager and senior project engineer, and construction supervisor.
- Review plans, identify issues, and agree on action steps to overcome them.
- Receive hand marked updates from the construction supervisor.
- Review the actual percentage complete versus the planned percentage complete, percentage milestones met, and approximate costs (from the plan) to be passed to the cost engineer.

The planner should be seen as proactive not reactive, predicting the future issues and proposing solutions.

The planner's main skill is in the art of communicating with the project management team, cost engineer, construction supervisor, and contractor. A good planner is a good communicator. The questionnaire that covers the planning subject is as follow. For the answers, contact me and the website www.elreedyma.comli.com.

Quiz

1. After you have been assigned to a project, according to the schedule, 50 percent of the project should be completed. You discover that the project is running far behind schedule. The project will probably take double the time originally estimated by the previous project manager. On the other hand, you discover that upper management has been informed that the project is on schedule. What will be the BEST action?
   - Try to restructure the schedule to meet the project deadline.
   - Turn the project back to the previous project manager.
   - Report your assessment to upper management.
   - Move forward with the schedule as planned by the previous project manager and report at the first missed milestone.

2. You are working on a large construction project that is progressing within the schedule. Resource usage has remained steady and your boss has just awarded you a prize for your performance. One of your team members returns from a meeting with the customer and tells you the customer said he is not happy with the project progress. What is the FIRST action you should take?
   - Tell your manager.
   - Complete a team building exercise and invite the customer's representatives.
   - Change the schedule baseline.
   - Meet with the customer to uncover details.

3. You have just been assigned to take over a project that your management has told you is "out of control." When you asked your management what the problems were they had no specifics, but said that the project was behind schedule, over budget, and the client was dissatisfied. Which of the following should be of the MOST concern to you?
   - The project is over budget and behind schedule.
   - There is very little documentation related to the project.
   - The client is very dissatisfied with the project's progress.
   - Your management is looking for rapid and visible action on this project to rectify the problems.

4. During a meeting with some of the project stakeholders, you as a project manager were asked to add work to the project scope of work. You had access to correspondence about the project before the charter was signed and remember that the project sponsor specifically denied funding for the scope of work mentioned by these stakeholders.
   What is the best action to take?
   • Let the sponsor know of the stakeholders' request.
   • Evaluate the impact of adding the scope of work.
   • Tell the stakeholders the scope cannot be added.
   • Add the work if there is time available in the project schedule.

5. You have been working for eight months on a twelve month project time. The project is ahead of schedule when one of the functional managers tells you the resources committed to the project are no longer available. After investigating, you discover the company has just started another project and is using the resources committed to your project. You believe the new project is not critical, but the project manager is the son of a board member.
   What is the best action in this situation?
   • Determine when resources will become available.
   • Ask upper management to formally prioritize the projects.
   • Use the reserve to hire contractors to complete the work.
   • Negotiate a new schedule with the other project manager.

6. Maintenance and on-going operations are very important to projects and should:
   • be included as activities to be performed during the project closure phase.
   • have a separate phase in the project life cycle because a large portion of life cycle costs is devoted to maintenance and operations.
   • not be viewed as part of a project--a project is temporary with a definite beginning and end.
   • be viewed as a separate project.

7. A project manager must publish a project schedule. Activities, start/end times, and resources are identified. What should the project manager do next?
   • Distribute the project schedule according to the communications plan.
   • Confirm the availability of the resources.

- Refine the project plan to reflect more accurate costing information.
- Publish a Gantt chart illustrating the timeline.

8. You are a project manager and your project schedule is tight and in danger of falling behind when structure and piping leads to disrupting status meetings by arguing with each other.
   What action should you take?
   - Separate the two until the project is back on track.
   - Speak with each team member and give each a verbal warning.
   - Discuss the problem with the manager of the two team members.
   - Meet with both team members and their function manager to determine the source of conflict.

9. A project manager needs to determine the resources needed for the project. Select the primary tool from the following:
   - Work breakdown structure
   - Schedule
   - Expert advice from functional managers
   - Expert advice from management

10. Your role is a project manager for a large project. One of your key resources has started to do his task behind schedule and work quality is beginning to suffer as well. You are confident that this person is well aware of the work schedule and required quality specifications. What action should you take?
    - Report the problem to HR for corrective action.
    - Reassign some work to other team members until performance starts to improve.
    - Meet with the employee in private and try to determine the factors impacting performance.
    - Escalate the situation to the employee's functional manager and ask for assistance.

11. What leadership style should you employ during the first two weeks of project planning?
    - Coaching
    - Directing
    - Supporting
    - Facilitating

12. Which of the following are represented by a bar chart rather than network diagrams?
    - Logical relationships
    - Critical paths
    - Resource trade-off
    - Progress or status

13. You are finalizing the monthly projects status report due now to your manager when you discover that several project lead disciplines are not reporting actual hours spent on project tasks. Consequently, this results is skewed in project plan statistics.
    What is the MOST appropriate action to be taken?
    - Discuss the impacts of these actions with team member(s).
    - Report team member actions to the functional manager.
    - Continue reporting information as presented to you.
    - Provide accurate and truthful representations in all project reports.

14. An engineering office is giving you so much trouble that your time available allocated to the project has gone from 20 percent to over 80 percent for this small piece of the overall project. Most of the available engineering office's deliverables are late and inaccurate and you have little confidence in this company's ability to complete the project. What should you do?
    - Terminate the engineering office for convenience and hire another seller.
    - Assign a group within your team to meet with the engineering office and reassign project work so that the engineering office work is easier to accomplish.
    - Meet with the engineering office to discover the cause of the problem.
    - Provide some of your own staff to augment the engineering office's staff.

4

Project Cost Control

4.1 Introduction

The first question to define whether the project is successful or not is if the project finishes within the budget or not.

You may think you are a hero if you save 50 percent of the project budget, but really you are loser if you do not define the scope well, reduce the activity, do not reach the target, or do a very conservative cost estimate. The project manager will say, "I save your company money". No sir, you lose the company money because, in most cases, the company will keep the money for this project to have a continuous cash flow to avoid project stoppers. About 50 percent of the budget is keeping hold and not investing in other projects. You should have a wide vision.

The cost involves many subjects as we will discuss the cost of a project as a whole from a decision-making point of view.

In the start of a project, the cost estimate is the main target data to obtain in order to define the project budget. During project execution, we apply cost control methodology to control and follow up the project.

The cost estimate calculation is usually done in more than one phase during the project. In the initial phase of studies, as in the appraisal phase, the cost is calculated by a simple principle and the borders of right and wrong are very large. The accuracy of the cost estimate calculation increases with the forthcoming phases of the project.

The possibility of increasing the actual cost of the project decreases with time until reaching the end of the project, where we reach 100 percent of the cost of the project.

The phase of feasibility studies needs people with very high experience in similar projects to provide an accurate value to the cost. Therefore, it must be made through specialized expertise from those familiar with such projects. Residential building projects are different than factories and petroleum projects, all of which have distinctive characteristics and needs of their own.

To illustrate the steps in estimating the cost, we will use a traditional example of going to a friend to purchase land for a new villa. Your friend requires a rough budget cost estimate, but this will be very complicated to obtain, as you have no drawings, calculations, or concrete data to calculate cost.

So your estimate value will contain high uncertainty. Someone with experience from an engineering office can provide a budget, but he or she at least needs to know the number of floors and the location of the land and then he or she will have a cost as per (USD/m^2) for this land location, the number of floors, and so on.

The previous example has a very preliminary stage and it is commensurate with small projects such as building a home. In the case of large projects, a budget is required for this stage by making surveys and taking soil boreholes and then calculating the cost of the project. Then, the entire project will be studied again to obtain the profit from it after minor clarity.

In major projects, such as large-scale and high-cost industrial projects, especially where the equipment and machinery for production are unique, the estimated cost in the initial phase of studies could be within plus or minus 50 percent accuracy.

The difference between the expected and real value is very large, but when there is a similar project then the smaller will be the difference between the actual and estimate cost, but in industrial facilities a large proportion of the cost depends on the equipment and machinery used in manufacturing.

On the other hand, there are some projects, such as the replacement and renovation of residential or industrial facilities, that have a very low percentage of accuracy and will also be about plus or minus 50 percent.

After finishing the FEED engineering, the estimate cost accuracy will be about plus or minus 30 percent.

After finishing the detail construction drawing for the whole project, the estimate cost will be calculated based on the quantity and will provide an approximate cost for each item. Then, we have an estimate cost for the project with a predicting, small deviation nearer to the actual and accepted accuracy, which is about plus or minus 15 percent in this stage.

After determining the cost of the project, the project has a budget that can run the project. After a while, you discover that the cost may be over 15 percent and the project may completely or partially stop. On the other hand, if the cost is less than the budget lower and less than 15 percent, you could book the amount of money you don't use and see whether or not the owner can invest it in another projects. Therefore, the cost estimate is very critical and vital to project success.

4.2   Cost Types

For the owner, there are different types of costs for construction projects, including the cost of assets and the capital cost, which include the cost of the initial composition of the facilities of the project. That cost is described as follows:

- Cost of land and property registration procedures
- Planning and feasibility studies
- Engineering activities and studies
- Construction materials, equipment, and supervision on site
- Insurance and taxes during the project
- The cost of the owner's office
- The cost of other equipment that is not used in construction, such as private cars to transport owner engineers
- Inspections and tests

The cost of maintenance and operation in each year of the life of the project includes the following:

- Leasing land
- Employment and labor wage
- Materials required for maintenance, repairs, and annual renewal
- Taxes and insurance
- Other costs of the owner

The cost values of each of the preceding items vary according to the type, size, and location of the project and the structure of the organization presence for other considerations.

We should not forget that the owner's goal is to reduce the total cost of the project to be consistent with the objective of the investments.

The highest value in the project cost is the cost of construction in the case of real estate and building structure. But, in the case of industrial building and the petrochemical industry, the cost of civil and structure work is almost small relative to other mechanical and electrical equipment. For example, the cost of a concrete foundation for a power turbine may cost 30,000 USD and the power turbine may cost more than five million USD. The other example is for nuclear plant or power generation projects.

When we calculate the cost from the viewpoint of the owner, it is very important to calculate the cost of operations and maintenance in each year of the life of the project for each of the alternatives available in the design and the cost of the life cycle of the project as a whole.

In calculating the estimated cost of the project to develop its own budget, we must point out the limits of deviation and the cost of risk or an unexpected event during the execution of the project. The percentage of risk has to be calculated for each item or in proportion of the total final cost. The calculation of the cost of risks depends on past experience and anticipated problems during the implementation of the project and increased costs of emergency often occur as a result of each of the following:

- Change in design
- Difference in the schedule and an increase in the time of the project
- Administrative changes such as increased salaries
- Special circumstances at the site, such as some unexpected obstacles or defects in the soil in some locations
- Special permits to work during the construction

4.2.1   Cost Estimate

A cost estimate is a prediction of the likely cost of the resources that will be required to complete all of the work of the project.

Cost estimating is done throughout the project. In the beginning of the project, proof of concept estimates must be done to allow the project to go on. An "order of magnitude" estimate is performed at this stage of the project. Order of magnitude estimates can have an accuracy of 50 to 100 percent. As the project progresses, more accurate estimates are required.

From company to company, the specified range of values for a given estimate may vary as well as the name that is used to describe it. For example, *conceptual estimates* are those that have an accuracy of 30 to 50 percent. *Preliminary estimates* are those that have an accuracy of 20 to 30 percent. *Definitive estimates* are those that have an accuracy of 15 to 20 percent. Finally, the *control estimate* of ten to fifteen percent is done. Early in the project, there is much uncertainty about what work is actually to be done in the project. There is no point in expending the effort to make a more accurate estimate than the accuracy needed at the particular stage that the project is in.

Types of estimates are described by Michel N.W. (2005). Several types of estimates are in common use. Depending on the accuracy required for the estimate and the cost and effort that can be expended, there are several choices.

4.2.1.1   Top-Down Estimates

Top-down estimates are used to estimate cost early in the project when information about the project is very limited. The term *top-down* comes from the idea that the estimate is made at the top level of the project. That is, the project itself is estimated with one single estimate. The advantage of this type of estimate is that it requires little effort and time to produce. The disadvantage is that the accuracy of the estimate is not high as it would be with a more detailed effort.

4.2.1.2   Bottom-Up Estimates

Bottom-up estimates are used when the project baselines are required or a control type of estimate is needed. These types of estimates are called "bottom-up" because they begin by estimating the details of the project and then summarizing the details into summary levels. The WBS can be used for this "roll up." The advantage of this kind of estimate is that it will produce accurate results. The accuracy of the bottom-up estimate depends on the level of detail that is considered. Statistically, convergence takes place as more and more detail is added. The disadvantage of this type of estimate is that the cost of doing detailed estimating is higher and the time to produce the estimate is considerably longer.

4.2.1.3   Analogous Estimates

Analogous estimates are a form of top-down estimates. This process uses the actual cost of previously completed projects to predict the cost of the project that is being estimated. Thus, there is an analogy between one project and another. If the project being used in the analogy and the project being estimated are very similar, the estimates could be quite accurate. If the projects are not very similar, then the estimates might not be very accurate at all. For example, a new technology in enhanced oil recovery is like a low salinity project, multiphase pumps, or subsea projects. The modules to be designed are very similar to modules that were used on another project, but they require more lines of code. The difficulty of the project is quite similar to the previous project. If the new project is 30 percent larger than the previous project, the analogy might predict a project cost 30 percent greater than that of the previous project.

4.2.1.4   Parametric Estimates

Parametric estimates are similar to analogous estimates in that they are also top-down estimates. Their inherent accuracy is no better or worse than analogous estimates.

The process of parametric estimating is accomplished by finding a parameter of the project being estimated that changes proportionately with project cost. Mathematically, a model is built based on one or more parameters. When the values of the parameters are entered into the model, the cost of the project results can be obtained.

Resource cost rates must be known for most types of estimates. This is the amount that things cost per unit. For example, gasoline has a unit cost of $1.95 per gallon, labor of a certain type has a cost of $200.00 per hour, and concrete has a cost of $100.00 per cubic meter. With these figures known, adjustments in the parameter will allow revising of the estimate.

If there is a close relationship between the parameters and cost and if the parameters are easy to quantify, the accuracy can be improved. If there are historical projects that are both more costly and less costly than the project being estimated and the parametric relationship is true for both of those historical projects, the estimating accuracy and the reliability of the parameter for this project will be better.

Multiple parameter estimates can be produced as well. In multiple parameter estimates, various weights are given to each parameter to allow for the calculation of cost by several parameters simultaneously.

For example, houses cost 150 dollars per square foot, software development cost is two dollars per line of code produced, an office building costs 260 dollars per square foot plus 55 dollars per cubic foot plus 2,000 dollars per acre of land, and so on.

The cost of the construction and installation is the biggest part of the total cost of the project and it is the largest share of the cost of the project. This cost is under control by the project manager and the project construction manager on site. The accuracy of calculating the estimated cost of construction is different from one stage to another. The more accurate the data is, the more accurate the calculation of the cost is.

The estimated cost of construction is calculated from more than one point of view, as from the owner perspective the cost estimated will be calculated based on the design and construction drawings. On the other

Table 4.1 Cost estimate procedure.

What	The function of the estimate is to forecast a cost for a specified scope of work to allow an accurate budget to be assembled for the business. Cost estimating in its definition is uncertain and the different classes of estimates give improved levels of accuracy as more project scope detail is defined. A work breakdown structure (WBS) should be built from which estimates, schedules, and cost control can be derived. Formal documents are produced for all levels of estimate. Appraise: Order of Magnitude (OOM) accuracy range of +/−50% Select: Class 3 accuracy range of +/−30% Define: Class 2 accuracy range of +/−15–20% Class 1 is +/−10% (rarely used).
Why	This is done to indicate to the business the predicted cost of the project, so that the project is financially viable and can be established.
How	The project leader and the asset development engineer develop a WBS from the SOR. This is then discussed with the project team to allow the relevant discipline engineers to have input in the estimate. An OOM estimate will normally be factored with estimates based on the known high-level scope and equipment definition. Class 3 estimates will be factored based on a developed scope and equipment definition, including indirect costs. Class 2 estimates will be built up from the developed scope of the project and will be fit for purpose to give the accuracy range.
When	All projects must have a Class 2 at the end of Define for business sanction. An estimate will be produced at the end of each development stage. The accuracy of the estimate will reflect that particular stage.
Who	The project leader is responsible and is supported by the estimator, discipline engineers, SPA, construction engineers, and the commissioning engineer.

side, the contractor will calculate the cost of construction in order to enter the tender and for this the contractor performs his calculation to qualify him to win the bidding. The third way is the control cost estimate, which is used by the owner to control costs, which will be discussed later.

The initial cost estimate is made after the preliminary studies of the project with an initial identification of industrial facilities, such as the number of pumps, air pumps, and compressors and the size of pipes and lines in diameter, and so on.

The cost is determined at this stage on the basis of previous experience of similar projects. For example, calculating the cost is as follows:

Example:

Calculate the cost for a project consisting of 3 pumps with 1200 HP, with pipeline 10 inches in diameter and a length of 15 miles.

The pump cost estimate = 700 $/HP.

Pump cost = 3 (700) (1200) =2.52 M$.

Cost estimate for onshore pipeline = 14000 $/inch/mile.

Pipeline cost = 14000 (10) (15) =2.1 M$.

Total cost = 2.52 M$+2.1 M$ = 4.73 M$.

Example:

Calculate the cost of one floor for a building with an area of 300 m² from reinforced concrete.

In this case, assume slab thickness is 250 mm for a slab beam and column as a practical estimate.

Approximate concrete quantity for slab and floor = 300 × 0.25=100 m³.

Assume the cost of concrete = 200 $/m³.

The concrete cost for one floor = $20,000.

The calculation of the concrete price is calculated after determining the type of concrete. The cost of reinforced concrete can be calculated by obtaining the following information:

- Quantity of steel per concrete cubic meter and the price of steel per ton
- Quantity of cement in concrete mix and the price of cement ton

- Quantity of coarse and fine aggregate and the price per cubic meter
- For ready mix, the price of each cubic meter by knowing the concrete grade, which is always 30–25 N/mm² in petro-chemical projects
- Cost of shattering, bending bars, pouring, and curing per cubic meter of concrete

For ready mix concrete, obtain the following information:

- Quantity of steel in the meter cube of concrete and the price of a ton of steel
- The price of concrete from the nearest ready mix location to the site and the concrete pump
- The cost of wooden form, which is usually a special strong form to the pump concrete
- The cost of steel fabrication
- The cost of the pouring process and curing

Calculate the approximate price for one meter cubed of reinforced concrete prepared onsite, noting that the approximate quantity of coarse aggregate is 0.8 m³ and fine aggregate is 0.4 m³ and this value is required to provide characteristic cube strength 25 N/mm²:

Steel cost = $1000/ton

Cement cost = $80/ton

Coarse aggregate cost = $5/m³

Sand cost = $1/m³

Steel quantity = 0.1 t/m³ as shown in Table (4.2)

Steel cost = 0.1 × 1000	= $100
Cement cost = 7/20 × 80	= $28
Coarse aggregate cost = 0.8 × 5	= $4
Sand cost = 0.4 × 1	= $1
Material total cost	= $133/m³ (as a material only)

Cost of fabrication of wood and steel and pouring = $30/m³

Total cost	= $163/m³

Note that in calculating the reinforced concrete cost estimate, the main item is the quantity of steel in concrete and it is different according to the

Table 4.2 Guide for estimating steel quantity in reinforced concrete.

Approximate Quantity of steel reinforcement in concrete, Kg/m³	Structure element type
90–100	Slab with beam
250	Flat slab
150–180	Hollow block slab
90–120	Columns
100–120	Isolated footing
200–300	Raft foundation

Table 4.3 Percentage of reinforced concrete building cost

Percentage from the total cost, %	Activity item
3	Design and site supervision
36	Concrete works
6	Masonry work
10	Sanitary and plumbing
45	Internal and external finishing

structure element. The following table is a guide for estimating the quantity of the steel reinforcement.

The data in Table (4.2) is considered a guideline and depends on the concrete characteristics, strength, and the member design, so it is an indicator for the quantity of steel reinforcement.

The percentage cost of the domestic and administration building will be a guide as shown in Table (4.3).

4.2.2   Steel Structure Cost Estimate

The cost estimate for the steel structure is significantly different than the calculation of the estimated cost of the reinforced concrete structure, as it needs a special design for the larger differences in the construction phase.

The most important part of the design and construction of the steel structure is the connection and it is known that the connection cost is about 50 percent of the whole building.

The following table defines the cost percentage for each part in a steel structure project.

Design	13%
Materials	38%
Fabrication	27%
Coating	10%
Erection	12%

From Figure (4.1), one can determine the proportions of the cost of main elements for constructing a steel structure. Every item has a percentage of the total cost.

We note from the above table that the connections, either through the use of welds or bolts, have the largest share in the process of preparing detailed drawings, where the most important and most dangerous phase is the accuracy of the details of the connection.

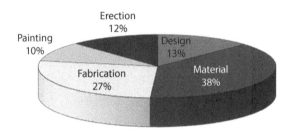

Figure 4.1 Percentage of cost for steel structures.

Table 4.4 Percentage of steel structure costs.

Item	Percentage from the total cost, %	Percentage of cost for connection, %	Percentage of the connection cost, %
Preliminary design	2	33	0.7
Final design	3	55	1.7
Detail drawings	8	77	6.2
Total design cost	13		8.6
Material	38	40	15.2
Fabrication	27	63	17
Painting	10	35	3.5
Erection	12	45	5.4
Total percentage	100		49.7

There is no doubt, also, that they are important in the cost of manufacturing and the installation phase, as the connection cost percentage is 63 percent and 45 percent, respectively. The connections are of great importance in the calculation of costs, as well as in the preparation of schedules.

4.2.3   Detailed Cost

The detailed cost estimate will be determined by the detailed construction drawings of the project and project specification.

The engineering office calculates the estimated cost through the experience of the contractors who carry out the construction of those projects with a cost estimate of others directly to the contractor as well as the expected profit. This depends on the engineering expertise of the office, in addition to the fact that the owner should add his cost for supervision.

The contractor will calculate the cost for the tender to be included in the bidding as the tendering cost estimate calculation is performed by the contractor and will be presented to the owner in the form of a bid or offer or as a start to negotiate the prices. Often, the contractor estimates the price with high accuracy in order to ensure that the work carried is out by the financial terms and he or she may use every method of calculating the estimated cost for construction.

Most of the construction companies have a procedure and scientific way to calculate the cost of construction of the project and to develop appropriate pricing of the tender.

There are some works and the facilities or equipment that will be delivered by the subcontractors. The subcontractors put the price of work and then the general contractor studies the offers from the subcontractors and chooses the best price. After that, the general contractor will add his percentage of indirect cost and profit.

It is included in the cost of supervision, non-profit, and expenses, in addition to the direct cost of others directly and it is conducted by subcontractors' offers, expense amounts, and construction procedures and steps.

4.2.4   Cost Estimate to Project Control

The cost estimate will be calculated to follow up the project during execution. The cost estimate can be obtained from the following information:

- Approximate budget
- The budget value after contract and before construction
- Estimated cost during construction

Both the owner and the contractor must have a baseline by which to control costs during project execution. The calculation of the detailed cost estimates is often used to determine an estimate of construction where the balance is sufficient to identify the elements of the entire project as a whole.

The contractor price, which was submitted by the tender with the time schedule, is determined by the estimated budget and is used to control costs during the implementation of the project.

The estimated cost of the budget will be updated periodically during the execution of the project to follow-up the cash flows along the project time period in an appropriate manner.

4.3   Economic Analysis to Project Cost

The project cost consists of two parts: direct and indirect cost. The direct cost is the cost of implementing the activities of the project set forth in the drawings and project documents. The cost of resources to be provided for the implementation of these activities are employment, raw materials, equipment, contractor fees, and other costs.

These costs do not enter the expenses related to the supervision of the implementation work. Indirect costs are expenses that cannot be avoided to supervise the works and facilitate the implementation and can also relate to the term "administrative and general expenses," which are usually fixed during the period of implementation of the project.

4.3.1   Work Breakdown Structure (WBS)

Most of the projects aim to achieve a certain degree of specific quality. This objective can be divided into a set of components that reflect the entire project. If the desired goal is to construct a gas plant, five main components can be identified, namely the following:

1. Public works site
2. Concrete structure
3. Mechanical
4. Electrical
5. Finishing works

After identifying the first level of the project, elements can consequently divide each of these components to a subset of components and can continue to divide the structure of the implementation of the work on successive levels of detail to achieve the level of the basic elements of project implementation. Therefore, importance is attached to the breakdown structure in making sure that all elements of the project have been taken into account.

The time spent in the preparation and processing of the WBS is, in return, its benefit, as it will be the main tool to avoid the forgetting or deleting of one of the important elements of the implementation of the project.

After determining the WBS, the work must be put in a particular method to identify the elements with a coding system.

4.3.2   Organization Breakdown Structure (OBS)

The Organization Breakdown Structure uses the same method for determining the levels of components of the project as the WBS. The system can be divided as well as the project management and organizational structure so that management knowledge can facilitate the identification of responsibilities within the project.

Moreover, the organization breakdown structure assists in surveying the work activities.

The organization should be known to everyone who is involved in the project in order to facilitate communication and define responsibility.

The organizational structure is usually followed by the project manager and a group of managers, each of whom oversees a portion of the project. For example, note that the electrical work supervisor will be responsible only for the activities required to implement the electrical work for the project.

It is clear how important the development of the project management's structure should be conducive to easy movement of information to and from the project manager to support his decision-making.

4.3.3   OBS/WBS Matrix

The organization breakdown structure and work breakdown structure matrix are shown in Table (4.5). The OBS is in the horizontal row and the vertical presents the organization.

From the above table, every person in the organization has a number of cells that present the work breakdown structure. For example, the project manager will do all the stages, so the cost will be for all these activities, the

Table 4.5 WBS/OBS matrix.

Civil work	Foundation work	Mechanical work	Electrical works	Design works	Planning work	
X	X	X	X	X	X	Project manager
					X	Planning supervisor
				X		Design supervisor
			X			Electric works supervisor
		X				Mechanical work supervisor
X	X					Civil work supervisor
	X					Concrete works supervisor

mechanical supervisor is allocated for mechanical work only, and the civil supervisor is allocated to civil work and foundation work.

4.3.4   Work Packages

The cost will be divided into work packages and this is the last level for the WBS that presents the project activity on which the time schedule and cost depend.

For every activity, it is required to define the following in order to calculate the cost:

- Execution time start and finish
- The resources to execute this activity
- The cost to execute the activity

Cost account (CA) is the cost budget for the activities that will be executed in this account. It will be a budget for everyone responsible in the organization and this direct cost is usually the material and labor cost to execute the activity.

For example, the concrete supervisor will be responsible for executing the following activities:

- The foundation with cost: $150,000
- The first floor with cost: $60,000
- The other floors: $400,000

So, in this case he or she will be responsible for three cost allocations with a total cost of $610,000.

Indirect cost, so far, is not taken into account in indirect expenses as the indirect expenses don't account in the work expenses or, to be more precise, in the execution of the WBS and, therefore, such costs are charged to the administrative management of the project or elements of project management.

For example, the concrete foreman in the previous example has a salary and consumes equipment and other expenses. These expenses cannot be added or charged to the specific activities, as he will supervise the foundation, columns, slabs, and others.

Those expenses must be paid on a regular basis during the execution of these activities. Also, any other expenses of supervision must be taken into account or any other administrative expenses such as offices for the engineers, technicians, computers, cars to transport engineers, and expenses and salaries of engineers and senior management.

For example, if concrete work will continue for a period of six months and indirect expenses as the former salary and miscellaneous expenses each month are $10,000, at the same time the cost of concrete materials and manufacturing from mixing until curing it cost around $440,000, it can all be calculated based on the quantity of concrete. Note that the former who is supervising this activity is included in the indirect cost by allocating for this activity for six months.

The total budget or estimated indirect costs is 6 × $10,000 = $60,000.

Thus, the total reinforcing concrete budget is as follows:

Direct cost = $440,000.

Indirect cost = $60,000.

Total cost = $500,000.

4.3.5 Cost Control

Cost control is very important in the management of projects, as they relate to the economics of the project as a whole, which is a key element in the success of any project.

The objective of cost control is a follow-up of what has been spent compared to what was already planned to be spent and to identify deviations, so as to do the appropriate action at an appropriate time. Therefore, cost control and the intervention of an ongoing process in the domain of the control of the project by the project manager who is directly responsible to define who is execute or supervise.

Calculating the actual cost should consider the different costs such as employment, materials, equipment, and sub-contractors and each calculation of the cost should be according to the specific document that all parties agreed on at the start of the project.

If the actual cost increases to more than the cost estimate, this will be due to one or more of the following reasons:

- The cost estimate is low.
- The circumstances of the project are not studied well.
- There is an increase in the prices of raw materials and labor during the project.
- There are climatic conditions and others that delay some of activities.
- There was a poor selection of equipment.
- There is inefficient supervision.

While it is difficult to correct the impact of the first four factors, there is always hope in improving the selection of the equipment and ensuring the department is aware and capable of choosing competent supervisors or increasing their capabilities.

The cost control process should be more than collecting data on the cost. The codification of data collection can be given a copy of the gain and loss after the implementation of the project. Cost control should help the project manager to analyze the performance rate for equipment productivities and manpower.

Reviewing the total spent on the project since the beginning of work until the date of the audit will present the situation of the project cost, which usually comes out of one of the following three cases:

It was exactly equal to the spending planned in accordance with the implementation plan of the project and estimated budget for this plan.

1. More has been spent than was planned according to the project's plan of implementation, which means an over expenditure or "cost overrun."

2. Less has been spent than the planned expenditure in accordance with the project's plan of implementation to end all activities, which entails a savings in spending or "cost underrun."

In general, over-expenditure is not desirable and must be prevented. The analysis of the causes must be identified so that it can be avoided in future. The savings in spending is desirable. However, this also requires searching about the causes of increasing cost as the main feature of the successful management of the project in the execution phase is the best way to perform a reduction in costs.

The following parameters are the main tools to control the cost:

- ACWP – actual cost of work performed
- BCWP – budget cost of work performed, also called earning value (EV)
- BCWS – budget cost of work scheduled
- BAC – budget at completion
- EAC – estimation at completion

To illustrate the above factors simply, assume that, in the phase of the engineering, CTR is planned to be done in 200 hours, the actual work will be 250 hours, and the work that is already done is 200 hours, the same as planned. One can see that what was done is equal to the plan. Assume the cost of one hour is $100.

The actual cost of work performed (ACWP) = $25,000.

The budget cost of work performed (BCWP) = $20,000.

The budget cost of work scheduled (BCWS) = $20,000.

Cost variation (CV) = BCWP – ACWP.

Schedule variance (SV) = BCWP – BCWS.

In the previous example the cost variance (CV) is equal to -$5,000.

Percentage of cost deviation = (ACWP – BCWP)/BCWP.

Schedule performance index (SI) = BCWP/BCWS.

A value higher than one represents an acceptable performance and a value less than one represents an unacceptable performance.

Cost performance index (CI) = BCWP/ACWP.

EAC =BAC/CI.

As stated previously, these factors must be calculated at regular intervals during project implementation and should, preferably, be compatible with the date of the month accounted by the company.

Monitoring these factors on a monthly basis will assist in evaluating the project and approximating information and the final cost of the project.

4.3.6 (S) Curve

The cost curve is called the (S) curve, whereas in all projects in the calculation of costs and distribution it to the Schedule, it takes the same form letter (S).

In the previous example, pouring concrete foundations is presented in the following figure: the first column represents the various activities and, on the other side, a Gantt chart was drawn and represents the cost of each item per day. For easy calculation assume that every activity cost is 1000 dollars per day.

This is done through the compilation of costs in the bottom row. With the first of the two cases, the situation is that each activity starts and finishes early and, in the second case, they start and finish late.

When we assign the cumulative cost curve and establish the relationship between cost and time, we will get the (S) curve, as in the following figure. We find the first curve in the case of an early implementation of the activities and the second curve as all activities that have been implemented in the latest time.

From the previous figure, we find that the cost budget is presented by the envelope curve. When the cost is higher than planned it should be in shape to do the work well; this pre-schedule completion is called "ahead of schedule."

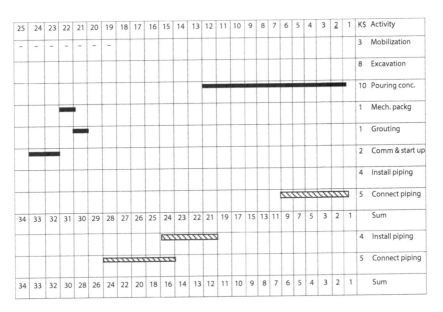

Figure 4.2 Distribution of cost on the activity.

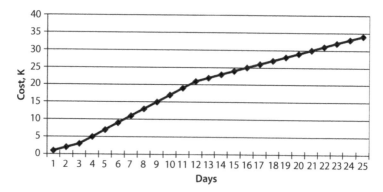

Figure 4.3 Cash flow in case of early dates.

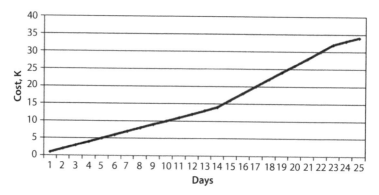

Figure 4.4 Cash flow in case of late date.

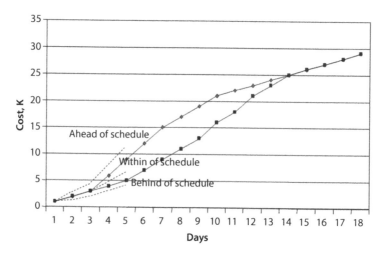

Figure 4.5 Cash flow envelope.

Table 4.6. Cost per month.

June	May	April	March	Feb.	Jan.	Plan
200	200	200	200	140	60	Work in month
1000	800	600	400	200	60	Cumulative
20000	20000	20000	20000	14000	6000	Cost/month
100000	80000	60000	40000	20000	6000	BCWS
Dec.	**Nov.**	**Oct.**	**Sep.**	**Aug.**	**July**	**Plan**
100	100	200	200	200	200	Work in month
2000	1900	1800	1600	1400	1200	Cumulative
100000	10000	20000	20000	20000	20000	Cost/month
200000	19000	18000	16000	14000	120000	BCWS

Table 4.7 Parameters calculation.

June	May	April	March	Feb.	Jan.	
20000	20000	20000	20000	14000	6000	Cost per month
100000	80000	60000	40000	20000	6000	BCWS
20000	20000	20000	20000	14000	4000	BCWP (EV)
20000	21000	23000	22000	14000	4000	ACWP
98000	78000	88000	38000	18000	4000	EV cumulative
104000	84000	63000	40000	18000	4000	ACWP cumulative
−2000	−2000	−1000	−1000	0	0	CV
0	0	0	−100	200	200	SV
0.942	0.929	0.921	0.95	1	1	Cost index
1	1	1	0.95	0.9	0.66	Schedule index
15000	15000	15000	15000	15000	15000	BAC
16000	16250	16333	16034	15000	15000	EAC

When the cost curve could prevent this cost within the envelope, this indicates that you are in accordance with the scheme called the time "within schedule."

Where the curve is less than planned, as shown in Figure (4.4), the situation is critical as the project is late and considered "behind schedule."

As an example, the following table gives the values of the cost of planning in an engineering project for a period of twelve months in accordance with the planning timetable explained above. In the project six months after the transaction, we will calculate the position of the crisis to see the project as presented in Table (4.6).

From the previous table calculate the cost control parameters after six months from start of the project.

From the parameters in Table (4.7), provide us tools to evaluate the project every month as follows:

January
The cost of execution is 300 less than the ACWP in this month, so the work is slow but the cost is acceptable.

The reason for this may be due to late hiring of new labours.

February
The work is progressing as planned, but work is still slow. Deal with this situation by letting them work on weekends to achieve the required time schedule.

March
During this month, the work is close to reaching the plan. Now the time schedule is not the only problem but also the cost increase.

April
This month the work was done more than it was planned. Now the work is going according to schedule and will be back to a normal mode of work, avoiding work on weekends.

May
The activity increased slightly more than planned, so we are going with the time schedule.

June
The work is progressing according to plan after six months of the project. Work is proceeding according to schedule, but there has been an increase in cost. It is expected that the budget at the end of the project will be about $16,000.

4.3.7   Engineering Cost Control

The most the project engineering phase has is ten percent of the cost of the total project. It is worth mentioning that most of the cost in design and drafting is around 58 percent of the total engineering cost. Table (2.8)

Table 4.8 Break down for engineering phase.

Activity	Percentage of the engineering cost
Design and drafting	57 %
Proposal	1%
Project Management	19%
Procurement	12%
Structural engineering	4%
Project Control, Estimating and Planning	7%

presents the breakdown of the engineering phase cost. The engineering contractors fee or the administration overhead and profit cost usually average five percent. It is shown in Table (2.8) that design and drafting are the most significant items of technical man hour cost, so, in most cases, it is easier to calculate the required design and drafting cost and then develop the other item as a percentage.

4.4   Cash Flow Calculation

The calculation of cash flow is presented in Chapter 3. Cash flow is the real movement of money to and from the project. The cash flow is positive if a company receives money. There is a negative flow of money if the company pays money and the difference between positive and negative cash flow is called net cash flow.

In the case of the contractor, the positive cash flow is the money received in invoices or monthly payments. The negative cash flow is the money paid toward labor salary, equipment, subcontractors, and other items during the construction. In any project, the contractor at some period will have negative cash flow, so he or she should provide money as an investment. In the case of increasing the net cash flow, this contract is self-financed.

In general, the construction companies will work by terms of contractual reservation in the case of a lack of funding in a time period during the project.

4.4.1   Project Cash Flow

The cash flows are calculated according to your role. In the case that you are the owner, you will pay the contractor and engineering offices

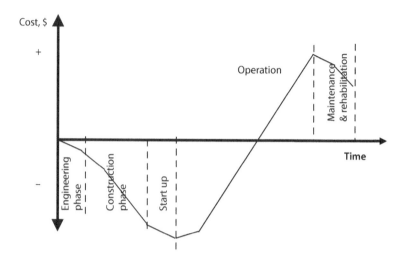

Figure 4.6 Cash flow along project lifetime.

throughout the project execution time. At the start of the operation of the project, the owner will gain money from the products, and, after a period of time the owner will obtain profits during the operation along the facilities' lifetime.

The process of cash flow during the period of the project has more than one phase, as shown in Figure (4.6), and is associated with the project plan and overall performance and cost of the project through the life cycle of the project overall.

The owner will pay the invoices for the preparation of feasibility studies and preliminary engineering studies in select stages during the first period of the project. When the project is in detailed design in the define phase, the project is increasing the number of team members on the detail design phase which correspondingly increases the cost.

The construction phase itself becomes a significant cost increase. In the construction phase, the rate of spending money is increased, as at this stage there will be large payments due to purchase of materials, equipment, and the contractor invoices and the maximum negative value at the end of the project and the start of the operational phase.

After operation begins and the product is sold, which increases with time, the negative cash flow will decrease until it becomes a positive cash flow.

It is worth mentioning that industrial projects, after a time period and according to the project design lifetime, require overhauls, maintenance, and rehabilitation that will decrease the profit as shown in Figure (4.6).

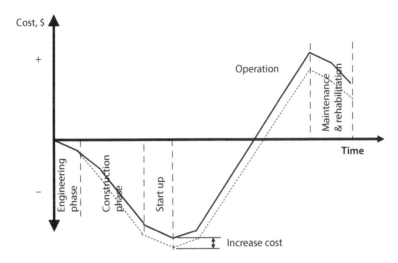

Figure 4.7 Effect of an increased cost on the whole project.

4.4.2   Impact on Increasing Cost

Sometimes the cost increases during the execution and there are many reasons for that. The most common cause is an increase in the price of materials from what was estimated before. There has been an increase in the foreign exchange rate, which often occurs in the machines that are imported from abroad. Another factor that increases cost is a difference in the quantities that have been executed and the quantities calculated from the drawings. An example that is common during excavation is that one can find different soil characteristics than what are in the soil report or the existence of problems in the soil were not taken into account, which requires restudying the soil and foundations with a change in the foundations design, which often causes an increase in costs. All of the above are project risks and if they occur they will eventually lead to an increase in project costs.

Figure (4.7) presents the impact of increases in the project cost with respect to the owner. From this figure, one can find that the period of time to start the profit gain from the project is at the start of positive cash flow. Moreover, the total income from the project at the end of the project lifetime is less than in the case of executing the project within the budget.

4.4.3   Project Late Impact

In the case of a project delay rather than a delay in the time schedule of the project, which occurs with poor project management, the planner and the

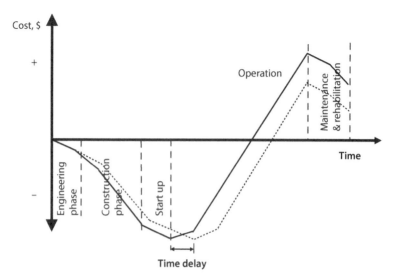

Figure 4.8 Effect of project delay on the whole project.

owner are often experienced in these matters and, therefore, those considerations are taken into account.

A delay may also occur because of a bad choice made by the supplier or the delay could be beyond their control.

No matter, the result is the same, and it is a delay in cash flow, as shown in Figure (4.8), resulting in an increased time period for the return. It is often the case that projects draw loans from banks as a result of this delay and increase the benefits. We find that the impact on profitability in end of the project lifetime may be more than in the case of only increased costs, as shown in Figure (4.8).

4.4.4   Impact of Operation Efficiency

The operation efficiency is the responsibility of the owner because errors occur during operation either as a result of faulty design or they don't meet the requirements of the owner and the required reliability. Errors usually lie between the owner and the engineering office.

The most common errors in industrial projects are due to the lack of choice of high quality equipment, which causes many problems and obstacles that affect the operation performance and, thereby, affect the overall revenue of the project. This is evident in Figure (4.9).

There is a last reason and an important element, which is the fact that operations need to have experienced and well trained staff, especially on the same type of operation.

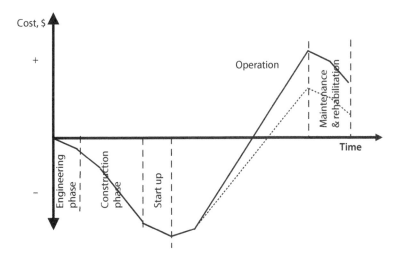

Figure 4.9 Effect of operation performance on the whole investment.

Any slack in the preparation of competent leaders or high professional operators will have a negative impact on production and, thereby, will reduce the overall revenue of the project.

The following is the cost control questioner that presents the actual cases we face in projects and the same questions you will answer in the PMP exam. Please try to solve for the best answer and contact me or present them on my website.

Quiz

1. The client's project manager asks you to provide a written cost estimate that is 30 percent higher than your estimate of the project's cost. He explains that the budgeting process requires managers to estimate pessimistically to ensure enough money is allocated for projects. What is the BEST way to handle this?
   - Add the 30 percent as a lump sum contingency fund to handle project risks.
   - Add the 30 percent to your cost estimate by spreading it evenly across all project tasks.
   - Create one cost baseline for budget allocation and a second one for the actual project plan.
   - Ask for information on risks that would cause your estimate to be too small.

2. You are in the middle of a new project when you discover that the previous project manager made a US $2,000,000 payment that was not approved in accordance with your company policies. Therefore, the project CPI is 1.2. What should you do?
   - Bury the cost in the largest cost center available.
   - Put the payment in an escrow account.
   - Contact your manager.
   - Ignore the payment.

3. A project manager for an offshore project is unsure how much cost contingency to add to the project. There is a 50 percent chance of a weather delay causing an impact of US $100,000 and a 20 percent chance of a delay in the testing center with a US $20,000 impact. How much should the cost reserve be?
   - Less than $50,000
   - More than $120,000
   - Less than $20,000
   - More than $54,000

4. You've just completed the initiating phase of a small project and are moving into the planning phase, when a project stakeholder asks you for the project's budget and cost baseline. What should you tell her?
   - The project budget can be found in the project's charter, which has just been completed.
   - The project budget and baseline will not be finalized and accepted until the planning phase is completed.
   - The project plan will not contain the project's budget and baseline; this is a small project.
   - It is impossible to complete an estimate before the project plan is created.

5. An analysis shows that you will have a cost overrun at the end of the project. Which of the following should you do?
   - Evaluate options to crash or fast track the project and then evaluate options.
   - Meet with management to find out what to do.
   - Meet with the customer to look for costs to eliminate.
   - Add a reserve to the project.

6. A project is seriously delayed. Earned value analysis shows that the project needs to be completed ten percent faster than the work has been going. To get the project back on track, management wants to add ten people to a task currently assigned to one person. The project manager disagrees, noting that such an increase will not produce an increase in speed.
This is an example of:
- Law of Diminishing Returns.
- fast tracking.
- earned value.
- life cycle costing.

7. You are in a project startup. As a project manager you are invited to meetings with the project execution team. What should you include in these meetings that would have the biggest impact on the project?
- Review of the action item list
- Review of identified risks
- Assignment of tasks to team members
- Estimating costs

8. After analyzing the status of your project, you determine that the earned value is lower than the planned value. What should you expect as an outcome if this trend continues?
- The actual cost will be lower than planned.
- The estimate at completion will be lower than planned.
- The project will finish behind schedule.
- The project will finish below the original cost estimate.

9. A project manager has just been notified by the vendor that the cost increased.
What should you determine first as the project manager?
- If there is enough reserve to handle the change.
- If another vendor can provide it at the original cost.
- If another task can save money.
- If the task is on the critical path.

10. Which of the following sequences represents straight line depreciation?
- 200$, 200$, 200$
- 140$, 120$, 100$
- 160$, 120$, 100$
- 120$, 140$, 160$

5

Resource Hiring

5.1 Introduction

The main tools you should manage are your resources. If you are a project manager on the owner side, the resources are usually the manpower. If you are a project manager on the contractor side, you have to manage the manpower and the equipment as well.

The manpower will operate the equipment, so in this chapter we will focus on managing human resources through strong organization and communication and the way to guarantee the quality of the output from the worker.

This chapter illustrates the most popular type of organization, as one is usually working in one of these types, so you should know the points of weakness and strength in the type of organization you working in.

On the other hand, if you are in a client role and working with an engineering company or a contractor, you should also know the characteristics of their organization.

As an owner in small projects, the people who are working on the project are from the owner company organization and, in major projects, they can hire an expat from any country worldwide. On the other hand, for the engineering company and the contractor, most of the people are collected from other, different projects. All the parties of the project, the owner, engineering company, and contractor, should decide the shape of his or her organization and what will be the link with the head office.

5.2 Project Organization

First of all, we need to know and understand the three commonly used organizational forms and the possibility of using any of them to suit the circumstances of the organization in the head office. Therefore, we must point out the advantages and disadvantages of each of these organizational forms and discuss some important factors that one could use in choosing which organizations are to be used in the project. Then, to consider the possibility of merging, we must look at some of the basic forms and examine, briefly, overlaps resulting from the use of organizational structure unincorporated. We will discuss some of the details of the organization of the project team and characterization of the role of different project team members, as well as some behavioral problems faced by the project team.

In most cases, the project manager does not have a significant impact on the organizational form of the project. Senior management usually chooses the organizational form. This organization form affects the project manager performance and causes it to work very hard to extend the organizational structure of the project to match the target. The project or construction manager on site must understand the organization structure interactions and relationships and experienced project managers often re-formulate the organizational structure of the project to fit with their perception of what is best for the development of the project.

5.2.1 Types of Organizations

All types of organizations have their advantages and disadvantages. Therefore, they must be taken into account when choosing the appropriate organization of the project according to the nature of the project, its characteristics, and prior experience for projects undertaken by this company or other similar organizations.

We must have a full knowledge of these types of organizations and the advantages and drawbacks whether you belong to members of the clients for the work of this organization, are one of the directors of the project, or are a member of the project because you must know the gain and loss that could happen when working on this project. In addition, through advantages in every organization you have the motivation and driving forces, which will be used to move the individuals who work with you. Through the defense of those features, you have the motivation of the company's head office.

The types of organizations are as follows:

5.2.1.1   Project Organization as a Part of the Company

This organization is used, often, for small projects that are within the company itself, such as industrial projects that entail increasing the product line, expanding the plant, or creating a new production line.

In this case, define the departments that are responsible for the project. This organization is also used in a type of project that contains new technology, such as buying a new advanced machine in which the supervision of the project rests and operation will be with the engineering management responsibility.

In this project organization, one of the departments will manage the project. This department will have daily work performed by individuals and the department shall be responsible for the project and report to the senior management of the company. Now let us imagine what the advantages and disadvantages are to this type of organization.

The most important features can be summarized as follows:

- We can easily transport the personnel to and from the project, which helps to exchange experiences and transfer modern technology to the largest number of company labor.
- It is easy to utilize the employment.
- The individuals are working on the project in the same section of functionality that they normally work and, therefore, their presence in the project will enhance the promotion of individuals.
- Increasing the experience of individuals who work in the project will increase the benefit to each individual and the company as a whole.

The disadvantages to this type of organization are as follows:

- The department is in charge of the project and the daily routine must be taken into account to determine the time that will be taken to implement the project.
- Technical problems can be solved easily as they work on the same job function, but they usually avoid the administrative work.

- This organization, in most cases, will not be given responsibility for a specific individual, so that work will be the responsibility of the department as a whole. This means that it may be two persons working on the same activity in the project as maybe one of them busy in the department of regular activity or on vacation or the work may be reviewed by more than one person, which will cause a loss of accountability in the project.
- The motivation of a working group will be weak, as the senior management will consider it their usual routine work.

5.2.1.2 Separate Project Organization

This organizational structure functions is an independent unit separate from the company and its connectivity is to discuss the company and the numbers of periodic reports on the achievements that will be sent to the company.

This sort of organization is in industrial projects and supervises projects for a company that may not have many projects at the same time.

There are some disadvantages and advantages of this organization and the most important characteristics are as follows:

The project manager has full authority to manage the project and reports to the Executive Director in the organization.

- Everyone in the team directly reports to the project manager.
- A separate working group is given a sense of independence, which creates a high level of commitment and accuracy in work.
- In this organization, the understanding of work orders and the implementation are easy.
- There will be powerful, fast decision-making and vitality throughout the project.
- It reduces the communication between individuals in the project and the people who work in the company head office.

The disadvantages of this type of organization can be summarized as follows:

In the case of more than one project in the organization, each project should have teamwork on it, causing a duplication of effort and increased costs if we look at the individuals who carry out the services and administrative affairs, which would be repeated work in all projects.

- The less communication there is between the project and technical divisions, in the case of projects that need special technology, the more difficult it is to transfer expertise from the company organization to the project.
- The project group is considered individual and there is rebellion against the actions in the head office.
- The project has a specific time, objective, and target with separate project team members. Through this environment of work, there will be a growing sense of interdependence with the project, which means the relationship between these team members and their counterparts in the company head office will be cooled, but, on the other side, the individuals in the head office believe that the members of the project benefited greatly from working on the project, which generates some feelings of hatred.
- There is control of the project personnel, the fear of loss of administrative promotion, and upward mobility in the company head office.
- This kind of organization has defects that can be prevented if an appropriate project manager is selected.

5.2.1.3   Matrix Organization

The matrix organization is most common in the case of construction companies or consultancy firms that have more than one project running at the same time. The target of this type of organization is to be used for special projects working in the field of advanced technology, so that they

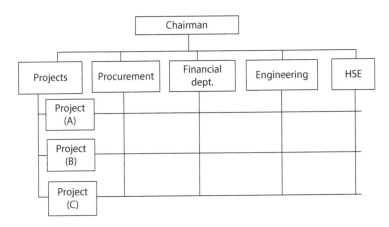

Figure 5.1 Matrix organization.

must be integrated disciplines for careers to work in implementing various projects.

The matrix organization will be functional posts in the parent organization, which consists of a set of integrated experiences that can be distributed on projects and sent back again.

This organization, although more commonly used in construction, can be summarized in the following points:

- You can take a technical engineer or any specialist at any time and when more than one project has been executed, the functional departments will have available expertise.
- There is no concern for the careers of individuals who work in the project upon completion of the project.
- There is strong communication between members of the project and between the functional departments in the main office.
- It has the flexibility as a separate project, but independent expertise is available as in the case that the organization is part of the organizational structure. The pain of any company is simply because it combines the advantages available to the two types of organizations.
- When there is more than one project, it can strike a balance between resources through good planning.

The disadvantages of this organization can be summarized as follows:

- The individuals in this organization are also scared of the end of the project, but not to the same degree as in the case of the independent project organization.
- The project manager controls the administration, while managers of the departments in the head office are managing the technical aspects only and this would cause confusion between the project manager and functional department head, as in some cases the technical decision may affect the project management from cost and time.
- Everyone has two managers.
- There is often overlap and conflict between the project manager and directors of different departments.
- A burden rests with the planning department in the head office in the distribution of resources commiserating with the nature of each project and its requirements.

Notes of such disadvantages can be avoided with an experienced project manager as well as select individuals who have experience commiserating with the project until they do obtain a balance between the requirements of the project manager and the functional department in the head office.

5.2.2   Selecting the Best Organization

It is difficult to identify a clear way of how to choose the type of organizational structure of the project, as the choice depends on the circumstances. There are some basics to be followed in selecting and designing the organizational structure, but there are no detailed steps that can be considered instructions to determining which type of organization is needed and how to build the project. All we can do in this regard is take into account the nature of the project and the characteristics of organizations that are available and the advantages and disadvantages of each. Then, we can make a comparison to find the best and most appropriate among these organizations.

In general, adjustment of functionality is valid in the case of projects that have the attention focused on the application of deep technology, more costs or delivery dates, responding rapidly to changes, or the adjustment of career preference for projects that require large investments in equipment or buildings and to the nature and functionality.

If the company deals in a large number of similar projects, such as construction projects, it is preferable to follow the organization of a single project. The single organization is generally appropriate to the projects that are unique, for one, and where privatization is high, functions individually, and requires careful control that does not comply with the department of a particular job.

When the project requires the integration of many functional departments and includes high-tech, but does not require that all the technicians are working, full-time professionals, the organization matrix is the best solution. But, it does mesh complexes and puts the project manager in difficult situations and, therefore, can be avoided when a structure is simple. When you choose the organizational structure for the project, the first problem is to determine the quality of work to be carried out. This is done through the initial selection of the objectives of the project and identifying the tasks necessary to implement each goal.

The following are a compilation of these functions in so-called "group work." Here, we could discuss the possibilities of the various organizational units at the plant in groups to do business within the project, also taking into account the restriction conditions inside and outside the head office.

So, by understanding the various organizations and their advantages and disadvantages, the senior management can choose an appropriate structure effective to execute the project.

In general, you can follow the following steps in choosing the organizational structure of the project:

1. Identify the goals of the project that need to be achieved.
2. Identify key tasks for each of these goals and identify functional units in the head office that can perform these tasks.
3. Figure the order of the main tasks and then configure the work package from them.
4. Determine which parts of the project will perform business groups and which will work with the other.
5. Make a list of the main characteristics of the project, for example the level of technology required, size and execution time expected, and, if there are no problems in allocating the resources, avoid the political problems occurring between the different functional departments who will be involved in the project. Finally, previous experience of the company will affect the project organization.
6. Based on full knowledge of the disadvantages and advantages of different organizations, one can select the appropriate organizational structure for the project.

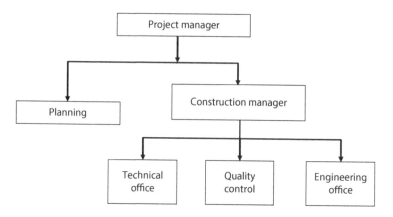

Figure 5.2 Organization chart on site.

5.3   Roles and Responsibilities of the Project Manager

The project manager is the key to the success of the project. His abilities can lead a team to success and not for a performance anomaly. Examples of the project manager's responsibilities are as follows:
Supervise the preparation and project plans.

- Define the relations between the different parties to execute the project.
- Take the necessary actions and procedures to obtain the project resources.
- Organize work between departments and supervise and coordinate among them.
- Follow up the project activities and make appropriate decisions to reform paths of execution.
- Monitor the costs and make decisions to ensure they match the plan costs.
- Take action to ensure the cash flows to and from the project is benefitting the project.

Ensure that the subcontractors are on the required level, and follow up their work.

- Preserve the rights of the company, which are represented in the contract, and oversee the administration of the contract.
- Send the deliverables for the project in a timely manner and review claims from the customer, which is the main way to know the degree of the project's success.
- Establish a system of reports that links and coordinates the project internally, with the functions departments, and externally, with the project owners, consultants, local authorities, sub-contractors, and suppliers.
- Attend the meetings of the project at a strategic level, in general, and executive level sometimes.
- Develop a policy to encourage employees in the project.
- The project manager is primarily responsible for achieving the project plan in terms of time, cost, and quality and, thus, should have powers to test assistants and subordinates and make all decisions within the project, including those not inconsistent with the policy of the company head office, regulations, and laws.

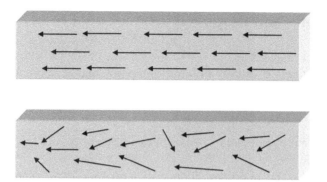

Figure 5.3 Effects of clarity and cloudy for objective, vision and mission to the project team.

- Ensure good construction management of the project on site in terms of administration, subsistence, regulating traffic, and securing the site and its employees against the risks. This will be achieved if the project manager is visiting the site on a regular basis. If the project manager was located away from the site, there has to be an appointed director of the project on site.
- Ensure that the materials and equipment work and are in compliance with specifications.

This makes it clear that the project manager is the conductor in the band. If there is any fault on his or her part to lead in the implementation of the business, the work will be done in a manner that is unsatisfactory to the client. In addition, he or she is responsible for clarifying the objectives of the project and what is required of them in accuracy and clarity.

The following figure illustrates the usefulness of the consensus of group work, setting goals as the business runs smoothly for the project as a whole. The second figure illustrates what happens with non-compliance and a lack of clarity. Time will be wasted and that would result in increased costs.

5.3.1   Project Manager as a Leader

A successful project manager has the ability to lead a team composed of individuals with different skill sets, personalities, and levels of experience. The team members may have also worked on different projects and in different organizational structures. As a project manager, you must overcome different cultural barriers and create a spirit of cooperation and coordination of efforts.

It is clear that the project manager must have some qualities and skills, which can be clarified and summarized as follows:

- Excellent communication skills
- Flexibility in work and acceptance of changes
- Training on the tools and techniques of project management
- The potential to direct every member of the project to achieve the objectives of the project
- Respect from senior management
- The ability to make quick decisions
- The ability to identify, analyze, and solve problems
- An entrepreneur mentality and work ethic and a governing of the general rules as role models
- Self-confidence
- Experience in procedures and project management tools
- Motivation to achieve success

There are two ways to manage individuals: through direct orders and sharp resolution of a centralized dictatorship or through democracy, which is the dialogue and discussion of ideas and analysis to issue commands. Each method has its advantages and disadvantages, but the project leader must determine how things will operate with the ability to change to another method easily.

Central decision or dictatorship is important in the case of a project that needs speed in implementation, depending on the importance and activity, which also generally depends on the nature of the project as a whole and must take into consideration the people who work with you and the nature of their personalities.

In many instances, this method is very successful, but it may fail with others, as well as by the general atmosphere of the project. For example, when there are good relations between individuals and there is a time for discussion, why not use the other way, a way of democracy to consider the views of individuals and make them reach the solution that you want. Control discussion, but let them feel that they are the owners of the idea. This would reach an amazing result, as the owner of the idea will try to forcefully demonstrate that it is a successful idea, and this is what serves the project as a whole and, thus, matches with the project goal.

5.4 Administrative Organization for Total Quality Management

In each company there must exist a team of total quality management that demonstrates in the project organization and the company organization as shown in the figures below. The following figures present various organizations and their relationship between different departments and the relations between the quality management team and other teams.

Figure 5.4 shows the administrative organization of the construction management on site, where we find quality control on the map of the administrative organization and its relations with the project manager and members of the team.

Figure 5.5 shows the administrative organization of the company. This company may be the office of the engineering company or contractor. It is clear that the presence of a special management quality system is based on follow-up and the existing manual of quality system with the presentation of the procedures and the required instructions to achieve the desired quality.

We also find that the quality general manager is in the same administration level with the project's general manager of the company with a direct relationship with chairman or to the Chief Executive Officer (CEO), which give the quality manager the required power to follow-up with the quality system in the whole organization.

It is, therefore, from the beginning when we want to make sure that the company has a system of quality assurance. We must look to the organizational structure and the site quality management of the overall

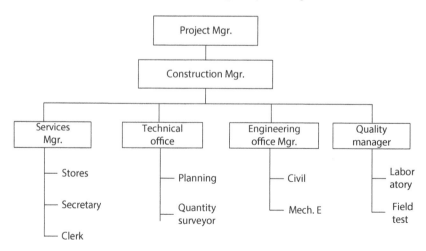

Figure 5.4 Sample of organization.

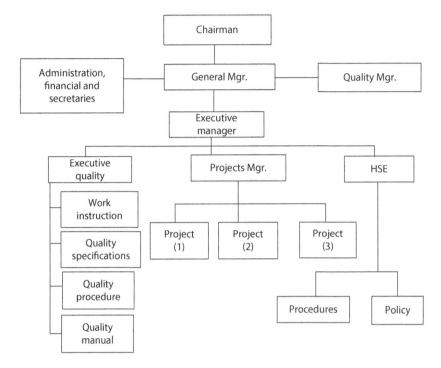

Figure 5.5 The head office organization.

organizational structure of the company, which can determine what the role of the quality system is and its impact within the company.

It is worth mentioning that when reviewing tenders and bids for any company, you have to carefully review the organizational structure through the quality of staff and their relationship with various departments and determine what the location and importance of quality is for the company.

From here we find that, in modern organizations, which take into account quality, you must see an existing slot to the quality general director, which has direct contact with the chairman of the company, as it gives him power and speed in decision-making.

The size of the quality team differs from one organization to another, depending on the size of project and the number of projects they are running.

5.5   Team Member Selection

The selection of the team members is the responsibility of the project manager. This selection is a very important step in starting the project. In some cases, some members will not be selected by the project manager and may be from your company or, in some cases, from the client if you are working in an engineering office or with a contractor. The competence of these persons will be a matter of luck, so the project manager should expect that and prepare to avoid this trap.

In most cases, the project team will be from different disciplines and different locations and are coming from different organizations with different skills.

In the matrix organization, the individual will have two direct managers: one who is responsible for technical aspects and the project manager. In some cases, a team member will be assigned to different projects and may be working part-time in your project.

When choosing the team members, you should interview and select every one individually. The team members may have worked together previously or may know each other. Some members may have not worked with you previously. So, your responsibility as project manager is to break the ice between all the team members and your challenge is to let them work together as one team.

The choice of individuals should be based on prior study for each individual and not by whether this individual is currently available. When selecting a team, you should ask yourself the following questions and be clear in answering each question:

- What is the experience related to the project?
- Does this individual have special skills suitable for the project?
- Does this individual have experience in similar projects?
- Has this individual ever worked in project teams before?
- Do he or she have technical information applicable to the requirements of the project?
- What is the department responsible for?
- Does he or she have a link with another project?
- When will this individual's association with this project end?
- What is the ratio of currency and participation in the project?
- What is the work load with him or her in the other project?
- Is it possible to reduce the workload by the excess of the project?
- Will he or she work from the beginning of the project until the end?
- Does he or she have a commitment to another project?
- Do this individual deal easily with people?

- Is there a record of his or her performance during another project?
- Is he or she interested to join in teamwork?
- Is this individual orderly, and does he or she know how to manage time?
- Does he or she take responsibilities very seriously?
- Is this individual an excellent player in teamwork?
- Is his or her handling of time commensurate with the tasks?
- What are his or her feelings toward the project manager's instructions (happy, sad, angry, etc....)?

Sometimes the answers to those questions are not easy, but to reach success everyone should be aware that each individual is important for the project.

The selection of a successful team does not support the selection of faithful functionality, only in the sense that we want civil engineers and we have civil engineers, but an individual who has joined the project should be capable of working in a teamwork environment. Also, the project manager should have confidence in good relations between the individuals. The selection must also be on the basis of the target you want to achieve and the required capacity of the team to complete the work properly under pressure.

5.6   Managing the Team

It should be clear that you have to face many difficulties when you start the composition of the group work. Difficulties often occur at the start of a project; to overcome this, you must enhance the interrelationship between the working group as soon as possible. These difficulties vary according to the size of team, as it is different for a team of five members and 50 members. You must pay attention to the following in order to overcome conflict and get the highest productivity from the team:

- Lack of clarity regarding responsibilities
- Lack of equal distribution of work between team members
- Lack of clarity in the allocation of work for each activity
- Lack of understanding of any of the stages of the project
- Overall objectives of the project are not clear
- Lack of trust between team members
- No continuous and strong contact between the team members

- Lack of guidance for the members
- No interest in the quality of work
- Objectives of individuals differ from the objective of the project.

Trying to overcome the above items will ensure a high probability of success for the team.

5.7   Allocate Resources to Project Plan

Now, we want to distribute the available resources at different stages of the project and the scheduled plan is obtained after the work of planning for the project as a whole is performed. Then, allocate the resources you have on the activities of the project. This will be explained in the following example.

5.7.1   Example

We will use the same example as discussed previously in Chapter 4 for time management. The project is to install a mechanical package and connect it with piping as shown in Table 5.1. From this table we can define the relationship between the project activities.

Figure 5.6 shows the critical path of this project through the project plan. Table 5.2 shows the early and late times for the beginning and end of each activity and the time which any activity can move within

Table 5.1   Example for foundation.

Item	Activity	Time (days)	Precedence activity
100	Mobilization	3	–
200	Excavation	8	100
300	Pouring concrete foundation and piping support	10	200,100
400	Install the piping	4	300,100
500	Install the mechanical package	1	300
600	Put the grouting	1	300, 500
700	Connect the piping	5	400,500
800	Commissioning and start up	2	700

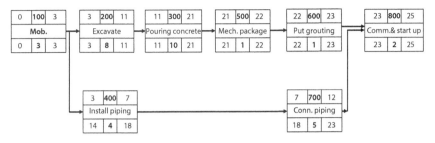

Figure 5.6 Preceding diagram define the critical path.

Table 5.2 Resources allocated to the activities.

No.	Activity	D	Earliest		Latest		Float		Resources per day	Total resources
			ES	EF	LS	LF	TF	FF		
100	Mobilization	3	0	3	0	3	0	0	4	12
200	Excavation	8	3	11	3	11	0	0	7	56
300	Pouring concrete foundation and piping support	10	11	21	11	21	0	0	4	40
400	Install piping	4	3	7	14	18	11	11	5	20
500	Install the mechanical package	1	21	22	21	22	0	0	4	4
600	Put the grouting	1	22	23	22	23	0	0	4	4
700	Connect the piping	5	7	12	18	23	11	11	5	25
800	Commissioning and start up	2	23	25	23	25	0	0	6	12
Total projects man days										173

itself and the overall time. An activity can move through itself without affecting the overall time for the project and it is presented in the table by the total floating time (TF). The column of the resources shows the number of personnel required for each activity per day, while the last column shows the total number of days required for each activity. For example, in Table 5.2 excavation needs seven workers per day and work will last for eight days, so this activity needs 56 working days in order to be achieved.

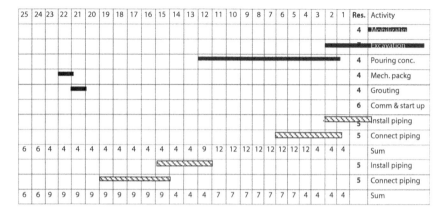

Figure 5.7

Figure 5.6 shows the distribution of the time duration of this project which is 25 days. When summing the total number of working days, as shown in Table 5.2, it will be 173 working days.

Figure 5.7 presents the time schedule and the number of employees. If the plan is presented by the early start, it means all activity will start in its first time to start. In the bottom of the schedule, the other option is to do all the activity in the latest start time, which means all the activity which is not on the critical path will be starting at its latest time. In our example here, the two activities are 400 and 700 for installing the piping and connecting the piping, respectively.

In the first case, the maximum number of persons on site will be twelve, whereas in the second option, for the latest start, the maximum will be nine persons. Therefore, a decision needs to be made on whether to use the total float or free float to achieve the project target. In normal cases, if the number of laborers will be less, the impact will be good, but if there is an increase in the laborers, the probability of incidents and injury will increase. We can look at it from another view as in remote areas, which are normal in the case of oil and gas projects, the accommodation will be difficult and expensive. So, if you maintain a minimum number of employees, it will have a good impact on the cost and, in some cases, it will be a constraint if the client provides accommodation for only ten persons per day.

In the case of the engineering design office, it is important to distribute the work load, as it will be very good to not hire engineers to work for part time for only six month, rather than hire the full time engineers. Distributing the work load by this philosophy is very beneficial

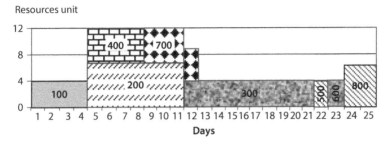

Figure 5.8 Resources distribution in first case.

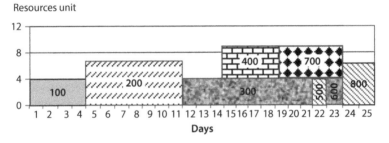

Figure 5.9 Resources distribution by latest plan.

Figures 5.8 and 5.9 represent the distribution of the resources in two cases and the harmony of resources distribution is visible if we use the latest start activity as discussed before.

It is worth mentioning here that, in the case of large projects, a meeting must be held between a representative of the owner, the consultant, and the contractor.

In this meeting, the contractor should define the personnel and equipment that will be found at the site in accordance with the scheduled plan. In this meeting, all the representative engineers should agree on the resources.

Noting that, each party will distribute resources according to its own benefits. As an example, the contractor in the negotiation will have had more than one reason for the distribution of resources, advertising and most other causes are undeclared as he may need some personnel for another project at a particular time or need to transport heavy equipment, such as cranes, to another site or other important resources. At the same time, he wants to delay some of the items to be cost assembled in another project and this will be studied in the chapter on costs.

From the owner point of view, it is important to complete all the items as soon as possible. The owner often does not want to consume all the time, which could be moved to minimize the risks to the project and use a higher employment rate in the start, which will lead to a less likely risk for project delay.

5.8   Relation Between Project Parties

Any large or small project depends on human resources, which are the main reasons for the success of any project or failure. In every phase of the project, there are teams that work either in parallel or in series. Different teams can work at the same time, teams can receive work from another team, and every team has its own organization and may have specific goals. The project manager is responsible for coordinating the relationships between all task forces and agencies that operate in the project.

The below figure shows the results of no relationships between the teams in the project. Assume you want to build a hammock in your backyard. This project is done in a number of stages, as shown in the figure, but in that case there is no connection or correlation between the teams working on the project.

In the stage of initial studies, the FEED engineering action team is in charge of studies. The initial proposal of the idea and the design team change it by hand. Then, the contractor will modify the idea once again according to its requirements or his or her point of view and will finally deliver the product to you.

5.9   Document and Information Transfer

The correspondence between the owner and the engineering office or the contractor is of paramount importance in any project transfer of correspondence and the speed and accuracy of response from the right person represents the most important thing to customer satisfaction. As we have stated before, all three parties are a customer and a client at the same time, so the movement of documents among the parties is extremely important for each one.

The following figures present an example of documents movement, but they vary from company to company. Figure 5.10 shows an example of the movement of correspondence within one of the companies, which achieves the ISO specifications.

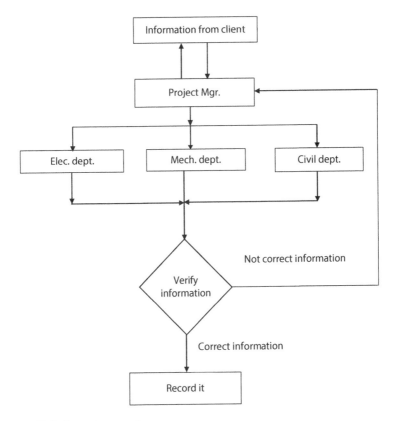

Figure 5.10 Information transfer.

We find that the movement of documents must be integrated and any document that enters the company should have a reaction that leads to a response to that document. In the end, the document will be saved. The complete integrated system is not allowed in a document transmitted and there is no response in any case.

5.10   Information Transfer

The transfer of information is of great importance, since this is a time of rapid information transfer and information technology, in particular with the development of the use of computers. To ensure that information will be transmitted fast and accurately, a strong procedure needs to be established in order to avoid any defect in the information transmission system.

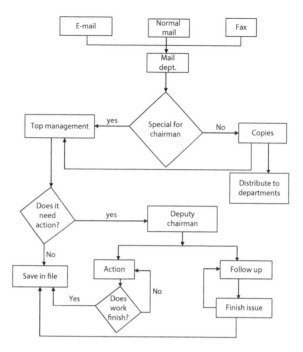

Figure 5.11 Document movement.

This shows the shape of the movement of information received by the project manager from the owner. The project manager receives information and distributes it to the heads of departments involved. The information will be reviewed and compared with previous projects and it will identify the degree of accuracy of such information.

If, after review, some errors in the information or inaccuracies are found, a report for the project manager is to be prepared, who, in turn, sends it to the client for review. If the information is sound and the project manager and director of the department are concerned to keep such information in a manner, to protect them from loss or damage which are registered on a special list.

5.11 Quality Control in the Design Phase

The purpose of this phase is to develop the required procedures to ensure that the design will be identical to the client's needs.

After the owner and engineering office on contract sign the final contract, the head office will select a project manager who will have direct responsibility for the project.

The main role of the office manager or general manager of the company is to verify customer satisfaction and follow up with the progress of work. Each project manager and general manager should determine the departments that will design work and should send letters to these departments to clarify what the project objectives are.

The project manager and the planning engineer have to specify the number of drawings required and the number of working hours and the work schedule of the project, which is determined by Form (2a).

The project manager will submit form (1a) to the heads of departments and the work file to the project to record data in a form (7a).

Before starting the design, the department head should determine the work plan using Form (3a), which contains a plan of action and the names of teams and estimates the number of working hours, the number of drawings required, and the foundations of design.

The responsibility of the project manager and planning engineer is to follow-up with the performance of all departments. The project manager and the planning engineer should start the preparation of the final time schedule after receiving schedules from all departments involved in this project.

One of the responsibilities of the project manager in the main office and the construction manager on site is to schedule meetings between the relevant departments for discussion or review of all departments and make sure that information is circulated between departments regularly and properly. It should fall to management to identify the engineer responsible for design, who gives approval of the drawings, and the Director of Administration should pursue his design through the timetable that has been prepared and sent to the Project Manager through the previous form (3a).

In the start-up phase, before the signing of the contract or in the tendering phase, the project manager should hold a meeting for all the disciplines, as with lead, civil, mechanical, piping, and others.

In most cases, the client sends information to the engineering office, usually the SOR, original drawings, or scope of work. The project manager then distributes a copy of these documents to all the discipline leads. In oil and gas projects, there should be good communication between the main departments, which are usually the civil and piping departments.

However, if the document is read by every discipline separately, every discipline may understand or interpret a paragraph differently. It is preferred that the project manager explains the project to the different disciplines and transfers the client requirements and expectations clearly to all the disciplines.

Engineering dept.: Date: Doc. No.:	1a	Engineering Company
Client: Phone: Fax: Email:		Project: Project No.: Project Mgr.: Date:

The required services and document:

Type	Description	Select
Tender evaluation		
Quantity surveying		
General plan		
Types of material		
Facilities work		
Report		
Specification		
Construction drawings		
Others		
Name: Date: Signature		

Engineering dept.: Date: Doc. No.:		2a	Engineering Company			
				Review meeting for client request:		
Client: Phone: Fax: Email:			Project: Project No.: Project Mgr.: Date:			
Activity	Inquiries	Time schedule	Man hours	No. of drawings	Description	Remarks
Design						
Tender evaluation						
Construction drawings						
Quantity surveying						
General plan						
Roads activity						
Specifications						
Other activity						
Name: Date: Signature						

Engineering dept.: Date: Doc. No.:		3a	Engineering Company		
Client: Phone: Fax: Email:			Project: Project No.: Project Mgr.: Date:		
Control Design:					
Sign.	End date	Start date	Work description	Engineering dept	No.
				.	
Date of first revision:					
Date of second revision:					
Date of final revision:					
Name: Date: Signature					

5.11.1   Inputs and Outputs of the Design Phase

Each department should determine with all team members what their role is, precisely, using Form (4a) to determine what is required from them in the design stage. We note that through this form input parameters in design for the project's general specifications and the type of structure are selected, the required structural analysis for if it is a preliminary or detail design is determined, and it is required to identify the required drawings' types as either construction drawings or just field sketches.

It also determines the size of the final drawing and notes that all the information is based on that which was defined by the owner and the engineering office through the SOR and BOD documents, which precisely reflect the demands of the owner.

Upon completion of the drawings, it is required that the final work is reviewed and team members have signed it, through Form (5a). Then comes the role of team quality assurance review, through the forms and making sure that each signature is in place in accordance with the responsibility of the person and that each engineer has passed all the steps required without neglecting any step. All this should be done before sending the drawings to the client.

We also note from this model any previous versions that have been prepared and recorded in history as well as a description of the detailed observations required by the owner.

It can also measure the difference between the estimated time to complete the drawings and the actual time spent, which is an indirect way of knowing the efficiency of departments and their time management.

5.11.2   Design Verification

The responsibility of the department head is to ensure that the design is in conformity with the standards and specifications agreed upon. In most instances, the department head has to return to some past projects in order to compare or request from another engineer to do some calculations to ensure the design. The registration is by Form (5a). If it is found that this item of the design is vague or unclear is registered through a Form (6a).

5.11.3   Change in the Design

There may be some deviation or change from what is stipulated in the contract, based on the owner requirement. The project manager should register the change, as shown in form (7a). Then, he or she should inform all the departments about the required change.

All changes are communicated to all departments through Form (8a). Send the documents by the change and make it clear that review of No. 1 is done in a different color.

Engineering dept.: Date: Doc. No.:	4a	Engineering Company	
Client: Phone: Fax: Email:		Project: Project No.: Project Mgr.: Date:	
Design Input:			
Option	Define requirement	Work description	
ACI,BS, EC		Specifications	
Final , preliminary		Structural analysis	
Feed, detail		Design	
No, summary, detail		Calculation sheet	
Steel, concrete, pre-stress concrete, others		Type of structure	
Sketches, construction drawings, as built		Drawings	
A0, A1, A4		Drawing size	
Start date:			
Finish Date::			
Name: Date: Signature:			

Engineering dept.: Date: Doc. No.:	5a	Engineering Company
Client: Phone: Fax: Email:		Project: Project No.: Project Mgr.: Date:

Review design and approval	
Date of Revision 1:	
Notes:	
Date notes completion:	
Date of Revision 2:	
Notes:	
Date notes completion:	
Date of final approval	
Remarks:	
Name: Date: Signature:	

Engineering dept.: Date: Doc. No.:	6a	Engineering Company
Client: Phone: Fax: Email:		Project: Project No.: Project Mgr.: Date:

Review design document

Item	Wrong	Right	Modifications
Technical data			
Dimensions			
General layout			
Error in Writing			
Other			

Name:
Date:
Signature:

Engineering dept.: Date: Doc. No.:	7a	Engineering Company
Client: Phone: Fax: Email:		Project: Project No.: Project Mgr.: Date:

Register project document

Export				Import				Register Doc.
Date	Review No.	Form	Doc. No.	Date	Review No.	Form	Doc. No.	

Name:
Date:
Signature:

5.11.4 Approval of the Design

After all departments agree on the design and send the drawings to the project, where they are reviewed according to Form (6a), make sure the design complies with the requirements of the owner and the specifications and standards that were agreed upon before it is sent to the owner through Form (9a).

When there are comments from the owner, the project manager will send the comments to the departments using Form (8a) and each department will review them and send them to the owner again. There is often more than one department responsible for this change, which can directly or indirectly affect from the main activity of the work before the amendment.

There are other departments, such as the cost control department, which are required to determine the change in costs as a result of the change in the original design and have the burden of negotiating the change in the price.

In general, the feasibility study may be affected by this change. Therefore, the department that prepared the feasibility study in the beginning has to determine the direct impact on the feasibility study, whether positive or negative as a result of the change in the design. This form is often completed by holding a meeting between all the departments in charge of the project with the proposed change of the owner and by taking suggestions from the various parties.

It is worth mentioning that, upon completion of the drawings, the work of all the changes and revisions are to be finally delivered. These final drawings must be stamped with red and marked as approved for construction.

Engineering dept.: Date: Doc. No.:	8a	Engineering Company	
Client: Phone: Fax: Email:		Project: Project No.: Project Mgr.: Date:	
Design modification			
Describe the modification:			
Department	Date	Select the department	Signature
Cost control			
Feasibility study			
Architect			
Structural			
mechanical			
electrical			
piping			
Evaluation			
contract			
other			
Name: Date: Signature:			

Engineering dept.: Date: Doc. No.:		9a	Engineering Company	
Client: Phone: Fax: Email:		Project: Project No.: Project Mgr.: Date:		
Cover transmittal				
Document type:				
Drawings	Report		Letter	
The objective from this document:				
Approval		Information		
No. of copies	Document name		Document number	Number
Receiving date: Signature:				

Quiz

1. Your organization is having a problem in the time management all of its projects. The CEO asks you to help senior management get a better understanding of the problems.
   What is the FIRST thing you should do?
   - Meet with individual project managers to get a better sense of what is happening.
   - Send a formal memo to all project managers requesting their project plans.
   - Meet with senior managers to help them develop a new tracking system for managing projects.
   - Review the project charters and Gantt charts for all projects.

2. During project planning in a matrix organization, the project manager determines that additional human resources are needed. From whom would he request these resources?
   - Project manager
   - Functional manager
   - Team
   - Project sponsor

3. A team member who doesn't have the required skills or knowledge was assigned to a team. Who is responsible for ensuring that he receives the proper training?
   - Sponsor
   - Functional manager
   - Project manager
   - Training coordinator

4. You work in a matrix organization when a team member comes to you to admit he is having trouble with his task. Although not yet in serious trouble, the team member admits he is uncertain of how to perform part of the work on the task. He suggests a training class available next week. Where should the cost of the training come from?
   - Switch to a trained resource to avoid the cost.
   - The human resource department budget
   - The team member's functional department budget
   - The project budget

5. What is the right way to overcome the cultural differences between employees in an international project?
   - Training through project management
   - Training for the different languages in the project
   - Training about different cultures and civilizations
   - Training about the differences in nationalities

6. You have just been informed that one of your team members has not been adequately trained to complete project tasks as assigned to him. How would you handle this situation?
   - Replace this team member with someone more qualified.
   - Request proper training be provided through the functional manager.
   - Revise the schedule to account for the decreased effectiveness of this resource.
   - Mentor this resource during the remainder of project duration.

7. A project manager has just been assigned a team that comes from many countries including Brazil, Japan, the US, and Britain. What is his or her BEST tool for success?
   - The Responsibility Assignment Matrix (RAM)
   - The teleconference
   - Team communication with the WBS
   - Communication and well developed people skills

8. One employee is three days late with a report. Five minutes before the meeting, where the topic of the report is to be discussed, he gives you the report. You notice that there are some serious errors in it. What will be your action?
   - Cancel the meeting and reschedule when the report is fixed.
   - Go to the meeting and tell the other attendees there are errors in the report.
   - Force the employee to do the presentation and remain silent as the other attendees find the errors.
   - Cancel the meeting and rewrite the report yourself.

9. The construction manager is away from the project and is replaced by a new one. The project manager meets with the replacement construction manager and his team. In this meeting, what is the first issue the project manager should start on in the meeting?
   - Introduce team members.
   - Communicate the objectives of the project.

- Clarify the authority.
- Create a communication plan.

10. There are over twenty stakeholders on your project. The project is running in another country with people from three countries as team members. Which of the following is the MOST important thing to keep in mind?
- The communication channels will be narrow.
- Many competing needs and objectives must be satisfied.
- There must be one sponsor from each country.
- Conflicts of interest must be disclosed.

11. A project manager has a problem with a team member's performance. Choose the best way of communication to address this problem:
- Formal written communication
- Formal verbal communication
- Informal written communication
- Informal verbal communication

12. You have just been assigned to a project that is in the middle of the execution phase.
What is the best way to control the project as the project manager?
- Use a combination of communication methods.
- Hold status meetings because they have worked best for you in the past.
- Refer to the Gantt chart weekly.
- Meet with management regularly.

13. What conflict resolution technique is a project manager using when he says, "I cannot deal with this issue now!"
- Problem solving
- Forcing
- Withdrawal
- Compromising

14. A team member complains to the project manager that another team member has once again failed to provide necessary information. The project manager meets with both team members to uncover the reason for the problem. This is an example of:
- Withdrawal.
- Confronting.
- Compromising.
- Smoothing.

15. You are now a project manager for an international project and you use people from different countries.
    What should you expect as a project manager?
    - Added costs due to shoddy or incomplete work
    - Language or cultural differences that preclude effective team work
    - Increased organizational planning and coordination activities
    - Team building activities become impractical and the cost is prohibitive

16. A senior engineer assigned to your project contacts you, trying to get off the team. He knows that an important project in his department is going to be approved and will take place at the same time as yours. He wants to work on the other project.
    What is the best action?
    - Release him from the team.
    - Talk to the functional manager about releasing him from the team.
    - Release him after he finds a suitable replacement.
    - Speak with the project sponsor about releasing him from the team.

17. Two lead processes and piping are having a big disagreement about how to accomplish a project from a technical point of view. The client is upset from that as it impacts time.
    What should you do as a project manager?
    - Make the decision.
    - Send the team members to their managers for advice on resolving the dispute.
    - Ask for a benchmark analysis.
    - Have the team members compromise.

18. Saying "Do the work because I have been put in charge!" is an example of what type of power?
    - Formal
    - Penalty
    - Effective
    - Expert

19. Which of the following is the BEST method to making a reward systems MOST effective?
    - Pay a large salary increase to the best workers.
    - Give the team a choice of rewards.
    - Make the link between performance and reward clear.
    - Present notifications of rewards within the company.

20. Your best professional structure engineer is a freelancer. Recently, you found out that he is working on a project in the evening for one of your competitors.
    What is the best action to take?
    - Replace him.
    - Get him to sign a nondisclosure agreement.
    - Inform him that you do not allow your contractors to work with your competition, and ask him to choose.
    - Limit his access to sensitive data.

.

6

Tendering, Bidding, and Contract Traps

6.1 Introduction

The contract is considered the backbone of any project. A contract with any and all parties is essential and vital.

The contract is an important factor for the success or failure of any project, depending on whether the contract has an impact on the behavior of the service provider and on your ability with your team to overcome the gaps in the contract. In this chapter we will clarify the types of contracts and tenders and the advantages and disadvantages of each type so that you can choose the best.

The most common case, in real business, is that you may have worked in the same place for a long time for a company that has a long presence in the market, which makes you feel that you will go through the tender and contracts procedures the same as your predecessors. You may move in their footsteps in preparing the contracts with the same ideas, but you will arrive at the same point that they did.

To be successful in your administrative activity you should think in the way that contracts were executed and how to resolve the gaps in these contacts. This can be identified after looking at the contractors that are working with you, reviewing their performance, and considering enhancements in their performance by changing the tender or contract procedures.

At the end of the close out reports of the project, which must be characterized by clarity and transparency so as to gain a benefit from previous projects, you should know the points of strengths and weaknesses, which lets you change your strategies in dealing with the various service providers until you reach a point of success that has not been reached before.

There has been great development in recent times at the global level in the fields of transportation and communication, which has had the greatest impact on the development of the industry where the ease and speed of communication and movement between countries and continents eventually lead to the free movement of trade.

Free trade between nations and between different continents has led to a fierce competition between all companies and international organizations that want to impose their control over global markets. Therefore, everyone tries to make products or services more unique than other competitors and, as a result of the intense competition strategies, become special in dealing with other firms.

The owner of the project may have a lot of offers from different countries. Therefore, the decision-making in a scientific manner is very important and, in most cases, there is a long distance between the client location and contractor engineering office or factory. This has led to the need for a system to guarantee the client's confidence in the product or service. This system is the ISO discussed previously. On the other hand, a good conclusion in the contract is one of the guarantees that will bring you assurance in the product quality while achieving the right price in a timely manner. Using a FIDIC contract will help with the success of the project and will help you achieve your goals.

6.2   Contracts

Contracts are one of the most important aspects in the construction industry where the relationships between the three parties are intertwined, as shown in Figure 6.1.

From this figure, one can conclude that there are two main contracts between the owner and the engineer and also with the contracts. On the

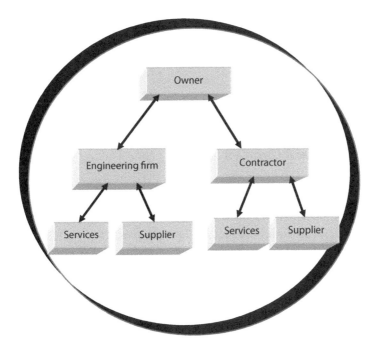

Figure 6.1 Contracts inside the project.

other side, the engineering firm or the contractors also have contracts to supply materials or provide services or subcontractors, and all these contracts will affect the project as a whole.

Therefore, the implementation of the project as a whole, in general, is linked to those contracts. For example, the services for an engineering firm stem from a company that maintains their computers. When there is no binding contract to achieve a fast response and quality in service of maintaining the computers, this will have a negative effect on the overall design time, which respectively affects the time of the final project or quality of the project. Therefore, any contract from any party has an impact on the whole project and a contract with the supplier of materials for the contractor is very critical, as any error will have a negative impact directly on the project time as a whole.

All contracts are important to the project and the goal of each contract should be clear to all.

The most risky contracts are the most important because they have the highest financial value in the whole project, which is the construction contract between the owner and the construction main contractor. The owner must determine the type of contract, how each contract is different depending on the nature of the project, and its objective. There are many types of contracts between the owner and the contractors. The most common types of contracts are the following contracts:

- Measured contract
- Lump-sum
- Cost plus contract

6.2.1   Measured Contract

This type of contract is most common in the construction field and has been used for a long time. The contract documents contain quantities of each item and each item is accurately described and has clear specifications. The contractor includes the price of each item and multiplies the price by quantity to determine the total cost of the item. The cost of compiling all the items is obtained in the total project cost.

The advantage of this method is that a shortage or increase in quantities can easily be known in addition to the price of this item, which is agreed upon in advance on the basis that is stipulated in the contract.

Recently, due to rising prices and the ongoing market changes, the presence of a fixed price throughout the project period may lead to damage for the contractor and, therefore, a high risk as there is no protection for the contractor during that contract.

The contractors have been reluctant or have refused to tender for contracts that have not provided some safeguard against rising costs. For those reasons, and to minimize economic risk, which might be exposed to it, there is an international agreement as a basis of business deals where, in the case of a project that exceeds its execution over twelve months, there will be a clause in the contract, thus allowing increasing prices in labor, materials, or tools. In European countries, a newsletter is published monthly to define the increase in prices through government institutions such as the Department of the Environment.

6.2.2   Lump Sum

Traditionally, this type of contract is used in projects with low cost. The contractor in this contract will be implementing the project at a fixed rate. Therefore, the drawings and specifications must be clear because this contract does not have a calculation of the quantities and, therefore, the amount of variation cost will be difficult to determine with any change in the site, which will cause protracted negotiations between the contractor and the owner, possibly causing disruption in the project as a whole.

This contract is now used as a turnkey contract to contain the design, procure, and execute at the same time (called EPC). It provides strong competition among contractors through good design, which reduces the total cost of the project and this will eventually have an economic return on the owner.

Preferably the contracts would be divided into the major items. For example, when the contract contains a purchase, supply, and installation of machine at the site, in this case, it can be divided into an item for civil works represented in the base concrete, an item for mechanical work, and item for the supply and installation of the machine, as well as an item of electrical activity. The contractor puts a price on each item of activity separately, so if there is an increase in the volume of concrete works, for example, we will negotiate the cost of the civil work only, which will greatly reduce the differences during the negotiations.

Some contracts also have an appendix that contains the price of labor per day or materials that we know by experience will cause an increase or decrease in this activity.

6.2.3   Cost Plus

In this contract, the contractor gets the real cost of construction with the addition of a fixed ratio throughout the project period. This figure contains the cost of supervision, the profit, and the cost of administration. This ratio is fixed for the duration of the project and when the price increase it does not become a problem. The disadvantage of this type of contract is that it will require a strong effort on the part of supervision and requires daily checking of the plant, labor, and materials.

The cost plus contract is normally used in projects that need to be executed urgently, are of particular importance, and are very small projects. They usually require delivery of materials in remote areas for urgent bases. This type of contract is common in oil and gas projects, as they require very urgent action. The cost of materials is known by the bill and the contactor obtains a fixed percentage that is agreed upon beforehand.

6.3   Contract Between the Owner and Engineering Office

This contract has some forms of fixed standards. In this contract, the description of the service is required from the engineering office. In most cases, the estimated value of this contract is within seven percent in small business and five percent in the large businesses in European and Arab countries. The Engineers Association has identified these percentages based on the size of the project and if the contract is for design only or design with supervision activities on site, there will be an additional fee for facilities, transportation, and other expenses during supervision.

In some projects, the supervision of the activity may be not a percentage of the project but the fixed cost for the project as a whole.

Too often in big projects the contract is based on the cost of man-hours and specifies the cost per hour of the lead engineer, senior, junior

Table 6.1   Cost, time, and resource estimate sheet.

Client		Project number	
Project title		CTR Number:	
CTR title:		Start Date	
Revision		End date	
Scope:			
In this section the scope of work will be written in summary but should be precise			
Assumption:			
The assumption that the designer will take into consideration that according to these paper the client will accept and review that.			
Inputs			
In this item will be the input data which will be the SOR from the client, soil data, survey maps, and other data that the client should deliver to the engineering firm.			
Deliverables:			
Doc: No.	Doc. Title		Doc. Type (report/drawings)
	Define the name of the document, which should be reviewed by the client carefully and should match the requirement of the SOR.		
Resources			
Resource	Hrs	rate	Total
Lead disciple			
Senior Engineer			
Junior engineer			
Senior drafting			
Drafting			
		Sum Total	

engineers, drafting, and others. In the project as a whole, the overall cost is determined by identifying the number of hours of each group's work and the price is specified hourly, containing any administrative expenses, insurance, taxes, and other expenses.

The benefit from the contract per man-hours comes from the request of the owner to change the part of the study design or add another part that is not in the scope of work that was delivered to the engineering office. The additional cost can be calculated easily because the office provides a number of hours for each member and the price is known, so this provides flexibility in the work as a whole.

6.4   The Importance of Contracts in Assuring the Quality of the Project

As discussed before, there are different types of contracts between the owner and the contractor as well as between the owner and the engineering office. Therefore, the contracts must be precisely defined by the nature of the project characteristics, which relies on the expertise of the project owner and its management. Bugs in the contracts will cause many problems and may be difficult to resolve, therefore wasting time and affecting the final cost of the project.

Therefore, it is recommended to review by knowing the objective of the project needs by highly experienced persons. A periodic specific review is required to gain a successful contract which will cover all aspects of the project without missing anything.

In general, the contract documents must contain the drawings and specifications for the materials, labor, and tools. The contracts must also determine the conditions of work on site and define the relationship between the owner, the contractor, the supervisory, and engineering facility in the contract's list of the quantities of each item and price.

There are some other basic items that are always overlooked by engineers with the vision of their importance, such as requirements of health, safety, and environment (HSE) to be followed, security of personnel and determining how to behave with the expenses in the case of injury or disease at the site, calculating taxes, defining courts that are referenced in the event of disagreement, and other administrative items that are not less important than the items that have been mentioned in engineering details.

A stable regime for international contracts is a contract called FIDIC and to maintain consistency there is a specific form for most items. FIDIC is the reference for international contracts and is used in major international projects.

6.5   Contracts in ISO

Review of contracts is set forth in ISO 9001 (ISO 9001 Section 4.3) and, in summary, the requirements for preparing and reviewing contracts in ISO are as follow:

- Contracts document
- Review contracts
- Procedures
- Requirements
- Capability of contractors or service provider

The contracts may contain a part of the ISO or refer to it. The contract document must contain a quality plan from the supplier or contractor, as well as plans to review the quality of the deliverables. It is strongly recommended that, before signing the contracts by the parties, it be reviewed closer and verified that all requirements are presented in the contract clearly. We must make sure that the supplier or contractor has the ability to perform the contract in the case of any more requirements in excess of the primary scope. The adding scope must be negotiated and incorporated into the final contract.

6.6   FIDIC Contracts

FIDIC contracts are the most popular contracts and are dealt with through international projects. FIDIC is the abbreviation for "Federation International Des Ingenieures Conseils." It means the "International Federation of Consulting Engineers." This union is comprised of assemblies of consulting engineers in different countries and was established in 1913 through the participation of three European societies and two architects of the French Society of consultant engineers (CICF), the Swiss architects consultant engineers (ASIC), and the Belgian Society of Consultant engineers (CICB).

The first conference was held in 1914 in the city of Berne. The consultant engineer has been defined in the conference as "the engineer who holds the scientific and technological knowledge and professional and scientific expertise, and who practiced law in his own name independent of any commercial enterprise or government for the client and does not receive any money except from his agent."

This union stopped working temporarily during World War II and then re-organized itself again in the 50s and 60s and emerged as a major development of the activity of FIDIC. It won international attention, strengthened its relations with the United Nations and World Bank, and lasted for some countries outside Europe such as Japan, America, and some countries of the Third World. Its membership includes about 65 associations of consulting engineers from around the world, including three Arab countries, Egypt, Tunisia, and Morocco.

The city of Lausanne in Switzerland is the headquarters of the Secretariat of the FIDIC. The most important activities carried out by FIDIC are the number of conditions typical of construction contracts and contracts covering different aspects of engineering, as there are books with red covers which focus on the contracts between the employer and the contractor for the civil works. A Yellow Book is special for the contract between the owner and the contractor for electrical and mechanical comprehensive installation. The Orange Book is for the turnkey design, while the White Paper is the terms of the employer with the consultant engineer.

The first edition of the book of general conditions for the implementation of civil construction work was published in 1957. Some items were added for the work of dredging in the second edition, which was published in 1969 and 1973. The review of the conditions were set forth in the contract with the World Bank and European Union contractors and after this review the World Bank approved the use of the terms of this contract in projects that the World Bank financed and rolled prints in 1987, 1988, and issued in 1992 and then issued as recently as 1996 in the book FIDIC. This book identifies the client as the employer and the Consulting Engineer, the contractor, and the definite relationships among them.

The book contains the general conditions of FIDIC's construction of civil works in the following:

Contents
Part I - General Conditions
Definitions and Interpretations
- 1-1 Definitions
- 1-2 Headings and marginal notes
- 1-3 Interpretation
- 1-4 Singular and plural
- 1-5 Notices, consents, approvals, certificates and determination

Engineer and Engineer's Representative
- 2-1 Duties of the engineer and authority
- 2-2 Representative of the engineer
- 2-3 The engineer's authority to delegate
- 2-4 The appointment of assistant
- 2-5 Written instructions
- 2-6 Engineer to act impartially

192 Project Management in the Oil and Gas Industry

Assignment and Subcontracting

3-1   Assignment of contract
4-1   Subcontracting
4-2   Assignment of subcontractor's obligation

Contract Documents

5-1   Language and law
5-2   Priority of contract documents
6-1   Custody and supply of drawings and documents
6-2   One copy of drawings to be kept on site
6-3   Disruption of progress (delay)
6-4   Delays and the cost of delay of drawings
6-5   Failure of the contractor to submit drawings
7-1   Supplementary drawings and instructions
7-2   Permanent works designed contractor
7-3   Responsibility unaffected by approvals

General Obligations

8-1   Contractor's general responsibilities
8-2   Site operations and methods of construction
9-1   Contract agreement
10-1   Performance security
10-2   Period of validity of performance security
10-3   Claims under the performance security
11-1   Inspection of site
12-1   Adverse physical obstructions or conditions
13-1   Carry out the works in accordance with the contract
14-1   Program to be submitted
14-2   Revised program
14-3   Submission of cash flow estimates
14-4   Contractor nor relived of his duties or responsibilities
15-1   Contractor's superintendence
16-1   Contractor 's employees
16-2   Engineer at liberty to object
17-1   Setting-out
18-1   Borings and exploratory excavation
19-1   Safety and security and protection of the environment
19-2   Responsibilities of the employer
20-1   Care of work
20-2   Responsibility to rectify loss and damage
20-3   Loss or damage arising from risk employer
20-4   Employer's risk

21-1 Insurance of works and the contractor equipment
21-3 Liability for the amounts not recovered
21-4 Exclusions
22-1 Damage to persons and property
22-2 Exceptions
22-3 Indemnity by the employer
23-1 Third party insurance (including the employer's property)
23-2 Minimum value for insurance
23-3 Cross liabilities
24-1 Accident or injuries to workmen
24-2 Insurance against accidents to workers
25-1 Evidence and terms of insurances
25-2 Adequacy of insurance
25-3 Remedy on contractor's failure to insure
26-1 Compliance with statues, regulations
27-1 Fossils
28-1 Patent rights
28-2 Royalties
29-1 Interference with traffic and adjoining properties
30-1 Avoidance of damage to roads
30-2 Transport of equipment by the contractor or temporary works
30-3 Transport of materials or plant
30-4 Waterborne traffic
31-1 Opportunities for other contractors
31-2 Facilities for other contractors
32-1 The contractor to keep site clear
33-1 Clearance of site on completion

Labor
34-1 Engagement staff and workers
35-1 Returns of labor and the contractor equipment

Materials, Plant, and Workmanship
36-1 Quality of materials, plant, and workmanship
36-2 Cost of the samples
36-3 Cost of tests
36-4 Cost of tests not provided for
36-5 Engineer's determination where tests not provided for
37-1 Inspection of operations
37-2 Inspection and testing
37-3 Inspection and test dates
37-4 Rejection

37-5 Independent inspection
38-1 Examination of work before covering up
38-2 Uncovering and making opening
39-1 Removal of improper work materials or plant
39-2 Default of contractor in compliance

Suspension

40-1 Suspension of work
40-2 Engineer's determination following suspension
40-3 Suspension lasting more than 84 days

Commencement and Delay

41-1 Commencement of works
42-1 Possession of site and access thereto
42-2 Failure to give possession
42-3 Rights of roads and facilities
43-1 Time for completion
44-1 Extension of time to the end the project
44-2 Contractor to provide notifications and detailed particulars
44-3 Interim determination of extension
45-1 Restrictions on working hours
46-1 Rate of progress
47-1 Liquidated damages for delay
47-2 Reduction of liquidated damage
48-1 Taking over certificate
48-2 Taking over of sections or parts
48-3 Substantial completion of parts
48-4 Surfaces requiring reinstatement

Liability for Defects

49-1 Defects liability period
49-2 Completion of the outstanding work and remedying defects
49-3 Costs of remedying defects
49-4 Contractor's failure to carry out instructions
50-1 Contractors to search

Alteration, Additions, and Omissions

51-1 Variations
51-2 Instructions for variation
52-1 Valuation of variation
52-2 Power of engineer to fix rate
52-3 Variations exceeding 15%
52-4 Workday

Procedure for Claims

Contractor's Equipment, Materials, and Temporary Works

Measurements

Provisional Sums

Nominated Subcontractors

Certificates and Payments

Changes in Cost Legislation
70-1   Increase or decrease of cost
70-2   Subsequent legislation

Currency and Rates of Exchange
71-1   Currency restrictions
72-1   Rates of exchange
72-2   Currency proportions
72-3   Currencies of payment for provisional sums

6.7   General Conditions in the Contracts

General requirements in the contracts must now be determined accurately and are often made by someone of legal or administrative studies while outlining the plan and the technical aspects of work for the engineer. But, in the end, the project engineer is responsible for any problem or error during execution, both technical or administrative. Therefore, you should review the contract and general conditions carefully. The following list must be reviewed and all items that have been drafted by the contract should be confirmed. This list is also used to terminate the contract, to make sure that the contract contains all the essential items, and to ensure that nothing has been forgotten.

List of General Conditions

Contract
☐ Definitions   ☐ Laws and language   ☐ Agreement document
☐ Clarification

Owner
☐ Owner responsibilities   ☐ Reach the site
☐ Deliver information and document

Owner Representative
☐ Engineer   ☐ Engineer representative   ☐ Roles and responsibility
☐ Requirements and conditions

Contractor
☐ General responsibilities   ☐ Guarantee
☐ Contractor Representative   ☐ Subcontractor

Engineering Design
☐ General responsibilities   ☐ Project specification   ☐ Drawings

Supervision and Labors
☐ Labor rules   ☐ Labors salary   ☐ Health and safety

Equipment and Materials
☐ Construction procedure   ☐ Tests

Time Plan
☐ Start time   ☐ End time   ☐ Performance rate oPenalty

Tests
☐ Contractor responsibilities   ☐ Testing procedure and condition

Receive Projects
☐ Receiving certificate   ☐ Operations manual to the owner

Fault Responsibility
☐ Cost of repair   ☐ Tests after repair

Contract Value and Payment
☐ Contract value   ☐ Down payment   ☐ Payment schedule

Change Order
☐ Right to change   ☐ Methods of change   ☐ Present change
☐ Value of change

Contractor Fault
☐ Excuse   ☐ Contract termination   ☐ Pay at completion

Insurance
☐ On design   ☐ Equipment   ☐ Labours oOthers

Force Major
☐ Definition   ☐ Effects
☐ Exemption from the delay and performance

Disputes and Arbitration
☐ Method claim   ☐ Payment of the claim   ☐ Arbitration
☐ Determine the courts in case of conflict
☐ The laws governing the conflict

6.8   Arbitration and the Arbitrator

Arbitration is defined as an agreement to present the dispute between two parties to a particular person or persons designated to provide a solution to this conflict instead of going to the courts.

In engineering projects, turning to arbitration as the dispute between the owner (employer) and the contractor or the employer, consultant, has a special nature in most cases and requires special expertise, both technical and legal, as the general interest of all parties is to end dispute in the shortest possible time because an increase in time negatively affects the whole project.

Partial arbitration, when the problem is between the owner and the contractor, is raised and presented to the consultant engineer, who will act as a judge, and the technical and administrative arbitration will be done by him. This is the first step to stop the conflict between the owner and the contractor. So, in this case, a competent consultant engineer who has strong experience and the capability to solve these problems will be very essential.

FIDIC contracts play a vital role and give power to the consultant engineer when there is a requirement for the contractor. Therefore, the first and most important step is to resolve the problem before it goes to the arbitration step.

Under the terms of FIDIC, the consultant engineer should be within 84 days from the date for receiving an order from any one of the parties in the conflict and the owner or the contractor should submit his decision by written document.

This decision is enforceable even though it may not be final and the party that does not apply the decision output considers the implementation of its contractual obligations a disruption. In particular, the contractor shall continue to work with due diligence, whether he or the owner requests to go through the arbitration procedure or not, until such time as amended by settlement agreement is in accordance with the decision of the arbitral tribunal. If any one of the parties does not desire to submit the dispute to a judge within 70 days from the date of the decision from the consultant engineer, this decision is considered final and the arbitral tribunal is to take into account the decision of the consultant engineer.

The contract must provide a way of arbitration and, often, the jury consists of three arbitrators chosen by each party and the two arbitrators agree on the third arbitrator, which is usually the president of the commission. International contracts must define the nationality of the third arbitrator as the decision will be crucial in the case of equality of votes in the arbitral tribunal. In the case of disputes in international contracts, the contract must define who will be the third arbitrator in the case that the parties fail to select who would be, it is better that it is the sponsor of the course of arbitration who choose the third arbitrator. Appointment authority comes from the management of arbitration in private institutions such as the International Chamber of Commerce President, judicial body, or person known for his or her integrity and it is usually recommend that the nationality of an appointment authority is different from the nationality of the two sides of the conflict.

6.9 Bids and Tenders

After defining the project and its work volume, collecting all drawings and specifications, selecting the type of contract, and determining the final form of the contract, we now enter the phase of complex administrative work. This phase is needed to understand all the laws of each country that has its own laws governing contracts and bids for projects, where it can reduce the manipulation and corruption and guarantee an honest and fair competition between contractors.

Most countries have laws that reduce construction in order to give some business to foreign companies and encourage local construction companies.

Understanding the laws governing the tendering and bidding procedures is one of the most important administrative steps that affects the whole project's primary objective to contract with the contractor who can provide the achievement of the project by the required quality at the lowest price.

Every country has its laws. Therefore, the international companies doing business in more than one country are responsible for a contracts department, which manages the contracts and tenders and has a strong knowledge of the laws governing contracts and tenders for each country.

The general framework of the different types of tender is fixed as well as the conditions to be followed and the characteristics of each type of bid, but there is a difference in the financial limits and some operational requirements.

There are four types of tender:

- Public (open) tender
- Limited tender
- Negotiated tender
- Direct order

6.9.1 Public (Open) Tender

The owner should prepare the service terms and conditions booklet and lists of works and accessories. Care must be taken into account in those documents and a copy of the contract and its specifications and financial, administrative, tax, insurance, and succession requirements must be kept. Translated brochures list the requirements and specifications and, in the case of external tenders, they indicate that Arab, French, Chinese, and other text is applicable in any disagreement or confusion in the content.

Open bidding should be advertised in newspapers and advertising should start in a timely manner. The advertisements should show the declaration submitted to tenders, the last date for submission, the work required, the value of the primary and final bond, the price of a copy of the tender conditions, and any other data the administration deems necessary for the work, whether external tenders must be advertised in the owner country and abroad also.

There must be at least thirty days for submission of bids in public tenders from the date of the first announcement of the auction. Licensed, competent authority may choose to shorten the period, but not less than fifteen days if necessary. Such failure does not apply to public tenders related to annual supplies, except in cases of extreme necessity dictated by the circumstances of the subject of the tender with the consent of the competent supreme authority, with the duration of validity of tenders from the date fixed for the opening of the envelopes to be decided in the auction and notification of acceptance before the expiry of the validity of the application of these tenders. If not that, the administrative body may request, in a timely manner, that bidders accept the validity of the extension as long as necessary.

Before the meeting of the opening the tenders envelopes, an employee should be assigned to receive bids that were received by the responsible department and to set out the relevant department for delivery for an immediate tender envelope which is coming in the morning directly to the open tender committee.

The chairman of the committee, on the day fixed for opening, the deadline for submission of tenders, is tasked with the following:

- Prove the case reported by the bidding after verifying the integrity of the seals.
- Establish the number of envelopes in minutes of meeting.
- Open the bids sequentially and for any open envelope the Chairman of the Commission puts on the envelope a serial number.
- Number the component of the tender and prove the number of such securities.
- Read the name of the bidder and the phrase to the audience of bidders or their representatives.
- Demonstrate the values of the tender and the value of insurance and avoid numbering.

The initial bond is not less than one percent of the tender value in the construction work. The successful bidder is to deposit in a period not exceeding ten days from the date of the next day by official letter with acknowledgment of receipt to accept the tender that will increase the bond equal to five percent of the value of construction work.

For international contractors outside the country, the bond should submit for a time period not more than twenty days. The concerning authority can extend the deadline for the filing of the final deposit not exceeding ten days.

This type of tender is usually determined by the government to enhance the flow of money inside the country, to develop new contractors, 032 and to enhance performance and support them financially.

6.9.2   Limited Tender

Limited tender is usually used in most special industrial projects, as there is a limited number of contractors or vendors working in industrial projects. It is very important that before registering the contractors on the company bidder list a pre-qualification assessment and audit for these companies should be performed as per ISO procedures and requirements.

In this case, the owner has a record of suppliers and contractors. That kind of tender calls for a specific number of contractors to be registered with it that are known to already have previous experience. This type of tender is carried out in most private companies and in some government departments to limit the participation by a number of suppliers and contractors involved, both within and outside the country.

In the procedure steps for the call for tender, the tendering and selection of the best bid uses the same method as public tenders.

More recently, as a result of the presence of the international network of information website, some international companies have been taken by the new launch tenders, especially in the case of limited tenders, which is well known in some minor companies.

The owner will make a website on the Internet, and the bidders for each company have the right to enter the page and change their prices. This method provides the ability to see the prices of the competitors. The advantage of this method is that there is a high transparency in the data and that all parties know why a company wins the bidding without getting into rumors and excitement, which was considered a type of negotiated tender, but through the Internet. On the other hand, it needs a precise description specific for each item.

This tender is commonly used in oil and gas companies, specifically for chemical tenders, as the specifications are very precise and defined well and there are only a few famous suppliers for these types of chemicals in the market worldwide. So it is very easy and helpful for the owner to obtain the best price.

6.9.3   Negotiated Tender

Negotiated tender is when the commission representative meets with the contractor(s) to reduce prices by bidding among them rather than dealing through direct negotiations.

We usually use this type of tendering in the following cases:

- Consultancy work, which requires technicians, specialists, or experts
- Things that only exist for one contractor
- Things monopolists manufacture or import
- Things that cannot be identified exactly
- Procurement, general contracting, and transportation services and the delivery of services, which are characterized by urgency
- Two competitive contractors provide the same price and, for urgency, we cannot redo the tender

Committees consist of technical members, financial members, and other legal bodies, such as the committee that is responsible for negotiating by reducing prices. This negotiation takes place in meetings between contractors and the bidding is done to reduce the price.

The contractors are invited to submit bids by sending the letters of acknowledgment of receipt containing the data to be mentioned in the announcement of public tenders and set out with these regulations to schedule the first meeting of the committee.

The invitation should be submitted to the largest possible number of operators in the type of activity, registered suppliers, and contractors.

The negotiation may also be advertised and the committee of negotiations will do the negotiation with all suppliers and contractors. Their discussion in public meetings is open to suppliers, contractors, or their representatives. The committee will then submit its recommendations to senior management, depending on what the competent committee was not authorized for by direct contracting.

6.9.4   Direct Order

This type of tender is usually performed in case of two situations: very urgent issues that can not wait for the normal bidding procedure or in the case that a product or service can be only be provided by one company. This is common in industrial projects or in the case that we need to develop a special engineering study and there is only one company that can do this engineering service.

In this case, the agreement will be direct to the vendor and negotiation about the price in this type of tender is rare.

6.9.5   Tender Technical Evaluation

Two envelopes will be submitted: one envelope provides the technical offer and the other contains the financial offer. The committee will open the envelope and the technical committee will evaluate the proposals technically. After you select the technically acceptable offers, the financial envelopes will be opened and evaluated and the lowest price and successful bidder are determined.

Evaluations of bids that are considered important and serious must be determined by specific criteria that assess the tender. Those standards are as follows:

Management system to the contractor company and its size

- The form of administrative organization at the site
- Previous experience of the contractor
- Quality standards criteria that the contractor applies
- Health, safety, and the environment (HSE) policy for the contractor and record of the latest report incidents
- Previous experience in the same project
- Equipment available
- The time schedule and the overall duration of the project completion

The above criteria are more or less important depending on the size and type of the project. Some projects have a specific nature. For example, for some projects the time factor is very important and an increase in time may affect the project. Hotel projects are an example where any delay reduces the number of days a hotel is open, which does not achieve any profit for the owner. Also, for oil companies, each day represents a loss of the number of barrels of oil a day. In those projects, considerable attention and weight is given to the schedule and time. In some projects, safety requirements may be the most important and, therefore, the safety criteria are the most important standards agreed upon between the members of the committee.

The best way to evaluate the bidding is by using the points system to determine the best technical offer. In this system, put a degree on each item. For example, twenty degrees is put on the duration of the project and the technical offer that gives the minimum period of time will take the twenty points, then a lesser degree is given to the other contractors who provide a longer execution time period. The company, which obtains the highest total is the winner, technically. In most cases every company has its own point system criteria, which is usually that any contractor has a total degree equal to or higher than 70 percent of the degrees or 60 percent or even 50 percent. If this acceptance percentage value is not in the company system, it will be decided through the technical committee members.

It is important that the table of points and the degree of each item, as far as accepting a 70 percent, for example, must be mentioned in the tender papers.

The below is an example for a tender evaluation point for a nuclear plant.

A. Experience

Table 6.2 Experience score.

No. of points	Item
50	1. General activity/standing of firm (year of establishment, number of employees, etc.) 2. Tender's past and recent experience and the consortium members (if any); they have participated in two nuclear power projects of 500 MW and above using water cooled reactor technology in the last five years as a minimum in two different locations. In addition, they have successfully completed three similar projects in the last 30 years, one of which is outside of home country. Tender and all the consortium members (if any) shall provide details of the required experience as well as its current workload.
100	• Site studies and evaluation
50	• Preparation of bid documents, invitation of bidding, bid evaluation and negotiation, and preparation of contract documents
100	• Project management and implementation includes commissioning
50	• Technology transfer and personnel training
350	Sub total

B. The Work Plan

Table 6.3 Work plan score.

No. of points	Item
50	1. Submittal of the tender in compliance with the terms and conditions listed in the tender document
50	2. Submittal of the tender in compliance with the technical requirement described in the tender document
75	3. Adequacy of the man months indicated in the proposal to execute the required services
50	4. Tender's proposed project time schedule and description of the approach to completeness of the work
25	5. Tender's localization plan to submit a complete program describing the components of the equipment that would be manufactured locally
50	6. Adequacy of the quality assurance program
100	7. Tender's organization chart and adequate manpower data of tender's personnel to be assigned during the execution of the work to meet the tender's schedule
400	Sub total

C. Financial Capability and Experience

Table 6.4 Score for financial capability and experience.

No. of points	Item
75	1. Tender's financial capability and audited financial data for the last three years
75	2. Tender to demonstrate its knowledge of the project international co-financing institutions procurement processes
150	Sub total

D. Industrial Safety

Table 6.5 HSE score.

No. of points	Item
20	1. Tender understands generally recognized industrial safety standards.
40	2. Tender's industrial safety and health performance has progressively improved over the last 3 (three) years.
40	3. Tender has submitted an acceptable industrial safety plan.
100	Sub total
1000	Total technical score

6.9.6   Commercial Evaluation

After the tender technical evaluation is finished the decision should now be made to award the project to the successful bidder. The commercial evaluation varies from one country to another due to special laws in government projects, but for private projects you have the right to make any system of evaluation match with your requirement. In some countries, for government projects you can choose the system of evaluation, but it should be stated on the tender package. The target is the same by any method that the owner requires to reach the best price.

Per our previous discussion, cost is an important factor, but so are other factors such as time and quality. If you choose the bidder who provides the lowest price, be careful that the lower price is not due to a mistake in the cost calculation or the scope not being clear. The impact on the project will be to the time or quality. I know that the experience and capability of the contractor will be covered by the technical evaluation, but you should remember that in technical evaluations we accept up to 70 percent, as previously mentioned. Unfortunately, most of the countries mention that they accept the lowest price.

6.9.6.1   Commercial Evaluation Methods

This phase is followed by the technical evaluation stage and is the most important stage in the evaluation process because it is the last stage, which will assign the project to the contractor directly.

Commercial evaluation methodology varies from one country to another and may also vary from project to project. Every law wants to reach the best price whatever the value of the tender, so in most laws in different countries determine the winner as who provides the lowest price.

On the other hand, there may be significant differences between the bidders that are not logical and this can cause problems during execution when assigned to the lowest price. This can be controlled directly through the accurate evaluation of the technical envelope when selecting the bidders company. This will be the lowest price method, which is good in the case of limited tender, but in case of open tender, the risk will increase as the contractor may be wrong in his price analysis or his estimation will have a very big impact on the project performance.

Some countries, such as Japan, have another method of commercial evaluation. They take the average of the numbers of the bids submitted after the deletion of the very highest bid and very lowest bid and the first bid value nearest to the medium will be the winner. For example, assume the technically accepted bids are 100 million, 500 million, 600 million, 700 million, and 1000 million, by taking the average of 500, 600, and 700 million only (omitting 100 and 1000 million) , the average is equal to 600 million. Based on that, the winner company provides an offer of 500 million.

For some military projects in some countries in the Arabian Gulf area, in the tender conditions it is stated that the winner will be just after the lowest price bid. For example, in the case of bids will be 1000, 700, 600, 500, and 100 million, the winner will be $500 million because it is the first value after the lowest offer. This method has an important advantage because every contractor knows that he will not succeed if the tender price was very low or high, so he will be very careful in pricing the items to match with the right number and won't reduce quality to be the lowest price.

Some countries and professional companies provide a relation between the technical evaluation point and the commercial value and this can be achieved by applying the following equation.

Tender will be ranked using the following criteria:

Total estimated price/Tender's technical evaluation score × 1000 LE

So, the lowest value will be the winner. For example, two companies have the same price of $1000 and the first one scores 1000 and the second one scores 700. The rank for the first one is 1.0 and the other is 1.4, so the winner will be the first one whether it is logical or not because the price is the same, but the technical evaluation is higher for the first one.

As another example, the first contractor price equals $800 thousand and the technical evaluation equals 700 points; the second contractor price is $900 thousand and the technical evaluation equals 1000 points.

For the lowest price technique, I will take the first contractor, which earns less technically. For this method the first contractor criteria = 800/700 = 1.14 and the second contractor = 900/1000 = 0.9, so the second contractor will be the winner. In this case, you pay less for a higher technical value.

6.10   Closeout Report

The closeout report will be in the final stage of any big project and, thus, the report will be prepared internally by the owner project team and should also be done by the contractor and engineering firm. This report is very important critical for future projects and is a main part for the cycle of continuous improvement. In most owner companies, after the commissioning and start up of the project, there will be a meeting for all the team members and a light workshop will present the advantages and disadvantages for any part of the project such as construction, engineering, procurement, HSE, contracts, planning, and other main activities. After this brainstorm meeting, collect all the data in a final report, which is the closeout report.

The characteristics of a final report are the following:

- It is prepared at the end of the project.
- It is prepared by the technical office, the planner, and costs control engineers.
- It is reviewed by the concerned departments and the final audit is done by the project manager.

Basic elements of the report include the following:

1. Introduction
2. Background on the project and its objectives
3. The budget allocated for the project and the actual cost
4. The difference between the estimated and actual cost allowed
5. Evaluation of the performance of contractors and suppliers
6. The final drawings correspond to the final situation on site (as built drawings)
7. Planning time at the beginning of the project has been modified in the time schedule

Table 6.6 Project closeout report preparation procedure.

What	This is a critical review of a project to assess both highlights and lowlights of project performance. This review should be used to ensure that lessons are learned and shared.
Why	To critically review project performance in all areas, i.e., schedule, cost, quality, etc. Good practices and outcomes should not be lost but should be shared with others. Less effective practices must be prevented.
How	For minor projects, the project leader should collate the report from core project members. For small/medium sized projects, a facilitated wash up session should be held to capture all aspects of project performance.
When	The report should be done in early operations to capture both good and bad issues.
Who	The project leader is responsible for the generation and communication of the report.

8. The reasons for completion of the project before or after the plan schedule
9. The change in the quantity rather than in the contracts and their impact on the time and costs
10. Recommendations:
    - Amendment to calculate the estimated cost
    - Modified schedule plan
    - Recommendations for the performance of contractors and suppliers
    - Recommendations for the operator

Quiz

1. You are reviewing bids from various contractors for work on your project. One of the bidding contractors has a history of delivering on time within budget and you have personally worked with this company successfully on previous engagements. You receive a call from the manager submitting the bid inquiring about how the process is going. He asks to have lunch with you to discuss the bid. What is the BEST response?
   - Do not mention the other bidders, but simply inform him that based on past experience, he has a good chance of winning the business.
   - Inform him that it would be inappropriate to discuss the matter at all and inform the customer or a team member of the conversation.
   - Inform him that it would not be appropriate to discuss the matter over the phone during business hours, but that an informal lunch discussion would be more appropriate.
   - Politely avoid continuing the conversation and disregard the bid.

2. Although your company is not the lowest bidder for a project, the client has come to expect good performance from your company and wants to award the contract to you. To win the contract, the client asks you to eliminate your project management costs. The client says that your company has good project processes and project controls unnecessarily inflate your costs. What should you do under these circumstances?
   - Eliminate your project management costs and rely on experience.

- Remove costs associated with project team communications, meetings, and customer reviews.
- Remove meeting costs but not the project manager's salary.
- Describe the costs incurred on past projects that did not use project management.

3. Bidders' conferences can have a negative effect on the project if the project manager does not remember to make sure:
   - All questions are put in writing and sent to all contractors.
   - All contractors get answers to their questions only.
   - To hold separate meetings with each bidder to ensure you receive proprietary data.
   - There is room in the meeting for all contractors.

4. An example of the contract price in a cost plus fixed fee contract is:
   - $2,0000 plus fee.
   - Costs, whatever they are, plus $20,000 as fee.
   - $20,000.
   - $250 per hour.

5. Which of the following factors can govern the project contract type?
   - How your company does business
   - How complete the scope of work is
   - Type of contract the law requires
   - Type of contract you have experience with

6. A company has just contracted with a well-known software developer to provide services during the planning and design phases of your project. Invoicing requirements were specifically defined within the contract, but expense limits were overlooked. As the project manager, which form of corrective action should you take?
   - Modify the terms of the contract.
   - Define acceptable limits to be adhered to.
   - Proceed in good faith.
   - Terminate the contract.

7. All of the following are generally part of the contract documents EXCEPT:
   - Proposal
   - Scope of work
   - Terms and conditions
   - Negotiation process

8. Your company is receiving a shipment of goods from the seller when you get a call from the contracting officer who tells you that the shipment does not meet the requirements of the contract. You look at the shipment yourself and determine that the shipment meets the needs of the project. What should you do?
   - Send the shipment back.
   - Accept the shipment.
   - Issue a change order to change the contract specifications.
   - Expect to receive a claim from the seller.

9. During the execution phase of the contract, the project manager should be concerned about conflict with the contract administrator because:
   - In many cases, the contract administrator is the only one who can change the contract.
   - The contract administrator is not interested in the contract.
   - The company favors the contract administrator rather than the project manager.
   - The contract is complex.

10. Your contract mentions that the maximum charge for services from the vendor will be US $40 thousand per month. However, the actual invoices have been US $90,000 for the past three months. Stopping the vendor's service will impact the project schedule. Under these circumstances, the BEST thing to do is to review the:
    - Contract change control system.
    - Scope change control system.
    - Performance reporting system.
    - Cost change control system.

11. An advantage of a fixed price contract for the owner is:
    - Cost risk is higher.
    - Cost risk is lower.
    - There is little risk.
    - Risk is shared by all parties.

12. The vendor on your project abruptly goes out of business. What should you do?
    - File for a portion of the company's assets.
    - Hire a new vendor immediately under a time and materials contract.
    - Terminate the project.
    - Terminate the contract.

13. You are in the process of having work crews dig a trench to lay fiber for a high-speed internet connection. All of the work permits have been obtained and funding has been approved. There have been several weather related delays, but due to perseverance of the entire team, the project is on time. It is the customer's responsibility to provide entrance into facilities so the connection into the building can be made. You discover the customer does not have adequate facilities and will not have them in time. What should you do?
   - Slow down the work, allowing the team time off but ensuring that you will be completed before the customer finishes their portion of the work.
   - Continue working according to your contract. Remind the customer both verbally and in writing of the customer's responsibilities. Provide the customer with an estimate of the impact if they do not meet their responsibilities.
   - Continue working with your company's portion of the work according to the contract. As a project manager, your job is done once this work is completed.
   - Stop all work and request that the customer contact you when they have fulfilled their responsibilities.

14. An IT manager says to you that he receives 35 new computers from the seller, but they were expecting only 30. In looking at the contract, you see that it says "seller to provide thirty (35) computers".
   - What should you do?
   - Call the seller and ask for clarification.
   - Return the five extra computers.
   - Make payment for the 35 computers.
   - Issue a change order through the contract manager.

15. From the contractor's point of view, the contract is considered closed when:
   - Scope of work is complete.
   - Lessons learned is complete.
   - Final payment is made.
   - The archives documents are completed.

16. You are working on your research and development project when your customer asks you to include a particular component in the project. You know this represents new work, and you do not have excess funds available. What should you do?
   - Delete another lower priority task to make more time and funds available.

- Use funds from the management reserve to cover the cost.
- Follow the contract change control process.
- Ask for more funds from the project sponsor.

17. Your company is very happy to work on this major new project. Noting that the contract is not yet signed, your management wants you to go ahead and begin to staff the project.
    - What should you do as the project manager?
    - Wait until the last minute to do so.
    - Ask the customer for a letter of intent.
    - Only start to collect resumes and not commit any funds.
    - Explain to management that this would not be a good idea at this point.

18. What is one of the KEY objectives during negotiations?
    - Obtain a fair and reasonable price.
    - Negotiate a price under the contractor's estimate.
    - Ensure that all project risks are thoroughly delineated.
    - Ensure that an effective communication plan is established.

19. Your company is purchasing the services of a consultant. You know one of the consulting companies interested in the work. What should you do?
    - Work hard to get the consulting company selected for the project.
    - Tell your manager and remove yourself from the selection committee.
    - Tell the people from the consulting company that you hope they get the work.
    - Keep the information to yourself.

7

New Approach in Managing Oil and Gas Projects

7.1 Introduction

The process of controlling and assuring the quality of a product is a critical subject now in most organizations and companies seeking to develop and create a privileged position at the local and global levels. So, essentially most of these companies are seeking to achieve total quality management (TQM).

In this regard, we will focus on quality systems in oil and gas projects and clarify the responsibility of all parties that share in the project and their responsibilities towards TQM through the management of the project.

All companies and organizations that have a target to grow their business in the open market now and a need to share in a suitable part of their market locally and internationally are going to apply the total quality management concept. In this chapter, it is important to focus on the way in which to apply the total quality management in oil and gas projects for engineering, contractors, and any supplier of the materials in these projects.

As mentioned before, the oil and gas business' concentration on low cost only is not good, but we usually focus on the project time to achieve the quality of the work and the materials that will not affect operations in future.

7.2   Quality System

Due to globalization, you can purchase any machine you want from any place worldwide, generate the need to have a system to let the client guarantee that the purchased machine will have the required quality, and deliver in the time agreed between the client and salesperson. So the client needs a system that gives him or her confidence in the products being purchased, thus decreasing the level of the risk in his or her project.

On the other hand, all the companies now for owner contractors, engineering, and vendors are multinational companies. So, the main office may be located in the U.S. and the other offices may be in the Middle East, Asia, or Europe. Therefore, can you imagine how controlling these offices gives a guarantee that the offices that are far away will deliver the product to the client in good quality and will protect the company reputation. So many are searching for a third party who can guarantee that the product will be delivered in a good manner.

For a long time, there was no third party capable of providing the confidence that a factory could provide good quality materials. Therefore, it becomes a real need to have specifications in order to achieve the quality assurance. Some specifications have been developed that control all steps of execution and manufacturing.

Work on these specifications began in the United Kingdom through the British Standards Institute, as it has been publishing a number of instructions on how to achieve BS4891 quality assurance. After a short time, a number of acceptable documents were made to meet the needs of the manufacturer or supplier.

From here, the specification BS5750 began and was published in 1979 through a series of specifications, which are guidance for internal quality management in the company as well as quality assurance of the product from outside the company. Quickly, standards became acceptable to the manufacturer, the supplier, and the customer. Then, the BS5750 standards became the benchmark for quality in the country.

At the same time, the U.S. was preparing a series of specifications, ANSI90, through the American National Standard Institute.

Some European countries also began preparing specifications in same direction. The British specifications are considered to be the base point for any European specifications.

7.3   ISO 9000

It's common to hear about ISO and its relation to quality. The International Organization for Standardization (ISO) was established in 1947 as an agent for the United Nations and it consists of representatives from 90 countries, sharing in the BSI and ANSI.

The activities of the ISO increase with time and there are many specifications published through the ISO. The specifications have widely spread due to the interest from manufacturers, their international agents, and customers. The manufacturer provides a product to give the customer satisfaction, which increases the production and sales.

Therefore, now machines and equipment are designed and manufactured in conformity with the international standards to guarantee that they will be acceptable for use in all countries, which increases the volume of sales and marketing of products.

The ISO 9000 specifications were released in 1987 and were very close to British standards BS5750 parts 1, 2, and 3. The same general arrangement of the parts and the ISO increased as a general guide to illustrate the basic concepts and some applications that can be used in a series of ISO 9000.

On December 10, 1987, the board of the European committee for standardization agreed to work on the specifications of ISO 9000. Also, it is formally considered a standard specification for European countries without amendments or modifications and it was published in the EN29000 1987. The official languages of the European standards are English, French, and German and this group agreed to publish and translate these specifications for every country based on its language. Then, the development of such standards was in 1994 when about 250 articles were modified. The articles often clarify the specifications and make it easy to read and with time the number of countries working with these specifications increased.

ISO 9000 has been divided for a number of parts to give details of quality assurance in the design, manufacturing, and acceptance of the final product. These parts are 9001, 9002, and 9003. These specifications can be used by the customer or the factory and can also be used as part of the contract between the seller and the customer. The ISO

9004 contains the basic rules for the development of the total Quality Management System according to the nature of the product, market, the factory, and available technology. Those specifications outline the definition of administrative regulations, the factory, and the customer, taking into consideration the requirements of society. Therefore, this part assigns the responsibilities of management and applies the quality policy.

7.4   Quality Management Requirements

The definition of quality management in ISO 9001 is an organizational structure of resources, activities, and responsibilities that provides us with procedures and means that make us trust in the ability of the company and any institution, in general, to achieve the requirements of quality.

7.4.1   Quality Manual

A company's quality manual is the formal record of the firm's quality management system. It can contain the following:

- A rule book by which an organization functions
- A source of information from which the client drives confidence
- A vehicle for auditing, reviewing, and evaluating the company's QMS
- A firm statement of the company policy towards Q.C.
- A quality assurance section and description of responsibilities

The source of information for the client should ensure the organization's ability to achieve quality. Identify responsibilities and relationships between individuals in the organization. This manual is the commander of the process, review, and evaluation of the quality system within the institution.

Therefore, this book must contain the following:

- Models of documents as well as models for the registration of the test results
- The necessary documents to determine how to follow-up on quality

7.4.2   Quality Plan

The quality plan contains steps to achieve quality in a practical way with the order of action steps to reach the required quality in the project.

The quality plan varies from one project to another according to the requirements of the contract with the owner or the client. In general, the contract provided is used to achieve quality in a certain way and select a particular goal that is needed for a plan of action for quality in order to reach what was requested by the client. The quality plan should include the resources he will use, different types of personnel, and equipment to achieve quality. It should also identify the responsibilities, methods, procedures, and work instructions in detail, illustrated with a program of testing and examination.

It is worth mentioning that the quality plan must be inflexible and cannot be edited, making it stable with time until the end of the project.

Some contracts require that the buyer, client, or owner projects have special requirements that are needed to achieve the final product, which must be clarified in the quality plan, detailing what steps should be taken to achieve what the client requires. This plan should be presented to the client to trust in the ability of the plant to achieve the quality of the desired product. The plan should include the following:

- All controls, processes, inspection equipment, manpower sources, and skills that a company must have to achieve the required quality
- Q.C. inspection and testing techniques that have been updated
- Any new measurement technique required to inspect the product
- No conflict between inspection and operation
- Standards of acceptability for all features and requirements that have been clearly recorded
- Compatibility of the design, manufacturing process, installation, inspection procedures, and applicable documentation have been assured well before production begins.

7.4.3   Quality Control

The quality control definition in ISO is a group of operations, activities, or tests that should be done in a definite way to achieve the required quality for the final products.

In construction projects, the final product is the building or structures, which should function properly. Therefore, in this case, the first step in quality control for getting the final product is to define the level of supervision in all the project phases to be sure that every part of the project is performed properly according to the required specifications. Try to ensure that the design, execution and the use of the buildings and structures are compatible with the project specifications.

Note that the quality control is responsible from any level from the manager to lower levels in the organization. In a practical case, the construction managers and the department head are responsible for quality control, but it should be clear that everyone is responsible for quality control, except the sponsor.

7.4.3.1   Why is Quality Control Important?

The improvement of quality provides many benefits. The use of quality control will lead to fewer mistakes by ensuring that work is being performed correctly. By eliminating the need for corrective rework, there will be a reduced waste from the project resources. Lower costs, higher productivity, and increased worker morale will then lead to a better competitive position for the company.

For example, consider two crews. Assume that each of the crews has the same crew size, skill level, and work activity. However, the first crew takes the benefit of having another person perform quality control duties. Therefore, if any defective work is built, it can be corrected before work proceeds any further. Any defect in the work made by the second crew will probably be discovered after the work is completed. This defect in the work will be torn down and corrected or ignored and left in place. Then, the latter choice will cause problems as construction progresses and will provide the owner with a degree of dissatisfaction. Customer dissatisfaction can cause the company to be removed from consideration for future construction projects or could require a costly correction due to the amount of work affected by the correction.

Additionally, defects are not free. The person who makes the mistake has taken money and the person who corrects the defective work will obtain money, too. Additional material and equipment costs will also apply to this correction process.

One example showing the effect of defective work is seen in the partial collapse of a parking garage in New York City. The absence of reinforcing steel in three out of six of the cast-in-place column haunches, which supported the main precast girders, was the cause of this accident. The project plans and the rebar shop drawings showed that reinforcing steel was to be installed at these locations but was accidentally left out. As a result, extra work had to be performed at the contractor's expense to correct the work and repair the damaged post.

Another major quality blunder occurred when constructing a shopping mall in Qatar. After pouring the concrete for the columns and the slab, they found around 40 percent of the columns had a strength lower than the allowable strength. So, due to the lack of concrete quality control on site and experience of the staff, this cost a lot of money to repair and delayed the whole project.

Quality is often "sacrificed" to save time and cut costs. However, quality does indeed save time and money. Nothing saves time and money more than doing the work the right way from the first time and eliminating the rework.

7.4.3.2   Submittal Data

The submittal data are usually the shop drawings, samples, performance data, and usually all the deliverable materials required test data results or Letters of Certification. The review of submittal data is one of the first steps in the quality control process. The information received from subcontractors and suppliers for items to be installed into the project must be verified to meet the standards set forth in the contract documents. Items such as dimensions (thickness, length, shape), ASTM standards, test reports, performance requirements, color, and coordination with other trades should be reviewed and verified carefully. Checking submittal information is important, especially when shop drawings are checked. Because the contract drawings do not provide enough detailed information to fabricate material, vendors and suppliers must make shop drawings. Materials that require shop drawings include concrete reinforcement, structural steel, cabinets/millwork, and elevators.

Basically, any item that is fabricated off-site is required to have a shop drawing in the submittal information. It is the information provided on the approved shop drawings that the fabricators use to "custom-make" their materials. Therefore, each item on the shop drawings must be verified against the contract plans and specifications. Once the submittal data is meticulously reviewed whether it is a set of shop drawings or some other form of submittal data, a determination on whether or not the data should be "approved" or "disapproved" must be made.

The General Contractor reviews submittals and they are given over to the consultant engineer for further review. If the data is "disapproved" or incomplete, the originator of the submittal data must resubmit correct or additional information. A submittal (set of submitted information for a proposed material/equipment) that completes the review process is used as the "template" by which the material is fabricated. Any mistakes not discovered in the submittal review process will lead to potential problems involving extra cost and additional time for correction.

The Kansas City Hyatt Regency walkway collapse in 1981 is an example of how a poor shop drawing review can lead to disastrous consequences. In this case, a change in the details of structural connections left unchecked during the submittal process, basically doubled the load on the fourth floor walkway connections.

This extra load on these connections led to the collapse of the fourth floor suspended walkway onto the second floor walkway and then onto the ground floor below. This disaster led to around 114 deaths and over 200 injuries.

7.4.3.3 How to Check Incoming Materials

Once submittal information is checked against the contract requirements and approved, it is filed for future reference. Many companies file submittals in reference numbers.

Verifying that incoming material meets contract requirements is accomplished by using the data shown in the submittal information. The information found on the delivery tickets or the manufacturer's information provided with the shipment is compared to the information given in an approved submittal. If all information is correct, then the material can be approved for off-loading to the storage site.

Take care that any "unapproved" material that is allowed to be stored on-site has the possibility of being included in the construction process and leads to rework or other corrective action. Therefore, it is important that each item coming onto the site is verified to comply with the contract requirements.

7.4.3.4 Methods of Laying Out and Checking Work

The layout of work and the verification of correct placement, orientation, and elevation of work are extremely important. Work that is not placed correctly will lead to an extra cost for rework. For example, the misplacement of anchor bolts for the foundation will lead to expensive correction work and delays.

In addition to checking work, the proper layout of work is also required. The required tools needed to perform this function include the use of a tape measure, plumb bob, carpenter's level, and a chalk box. Topics to discuss for the proper layout and checking of work include checking elevations at the height of concrete footing during placement and finishing the grade and floor. Methods to check for proper alignment of work in the field manufacturer's recommendations for the layout of certain items are windows, overhead door, and air-handling units.

Since quality control is the responsibility of everyone involved in the construction process, most of the engineers in construction positions will help to manage QC functions. Since it is not always clear what one needs to find in order to ensure a proper inspection, engineers should be instructed to watch for «key items" during inspection.

Example of QC Items for Steel Door and Frame Installation

1. When delivered to the site, each door and frame should be checked for damage.
2. Ensure proper size and gauge of doors.
3. Doors and frames must be stored off the ground in a place that protects them from the weather.
4. Do not stack doors or lay doors flat. This will cause doors to warp. Doors must be stacked on end of a carpet-covered racks or using other appropriate methods.
5. Check doors and frames for proper material, size, gauge, finish (satin, aluminum, milled), and anchorage requirements.
6. Verify door installation per door schedule shown in contract documents.
7. Fire-rated doors or frames must be used in fire-rated wall assemblies.
8. Fire-rated doors and frames must have a label attached or a certificate stating the fire-resistance rating.
9. Check for the proper location of the hinge side of the door and for proper swing of the door. (For example, per fire codes, the door swing for stairwells and other egress openings must open out, not into the stairwell.)
10. Door frames in masonry walls must be installed prior to starting masonry work (masonry must not be stepped back for future installation of doorframe).
11. Is the doorframe installation straight and plumb?
12. If wood blocking is required for doorframe installation, make sure this activity is completed during the construction of the wall.
13. There is a uniform clearance between the door and doorframe (usually 1/8").

14. Has adequate clearance been provided between the bottom of the door and the floor finish (carpet, tile) that will be installed?
15. Touch-up scratches and rust spots with approved paint primer.
16. Exterior doors must be insulated.
17. Check for weather-stripping requirements on exterior doors.
18. The intersection between the doorframe and wall should be caulked – check for missing caulking in hard-to-reach areas (for example, hinge-side of doorframe).

7.4.3.5 Material/Equipment Compliance Tests

Every project owner requires testing of materials and equipment prior to placement and after installation. Engineers should be familiar with testing methods, whether or not they will be performing the actual tests. Prior to beginning construction operations, a listing of each test that will be required should be made out.

This will serve as a checklist to be used by QC personnel. This testing checklist should list the type and frequency of testing required per each segment of work. Once tests have been performed, a test report documenting the results of the test should be kept on file or put into a "test report" folder for future reference. The following tests are typical tests that will be performed on the jobsite to ensure the quality of work is placed or completed.

7.4.3.5.1 Soils Testing

The foundation of a structure is responsible for transferring the loads from that structure into the ground below. The soil in this ground must be strong (dense) enough to stand with the loads that will be imposed. Additionally, the strength of soil must also be uniform to avoid any differential settlement in the structure, which can possibly cause structural and weatherproofing problems. In order to ensure that minimal settlement takes place in the building structure, the compaction of the soil must be verified. Each excavation or soil backfill operation must be checked to ensure compliance with the compaction requirements listed in the project specifications. These tests take place prior to starting any additional work, such as rebar placement.

7.4.3.5.2 Concrete Tests

There are two types of concrete tests that are used to evaluate concrete on the jobsite: the slump test and the concrete cylinder or cube test. The slump test, per ASTM C 143, determines whether the desired workability of the concrete has been achieved without making the concrete too wet.

7.4.3.5.3   Mortar Testing
The project specifications for mortar list that mortar must comply with either ASTM C270 or ASTM C780.

ASTM C270 states the required proportions of mortar ingredients (one part Type S masonry cement to three parts masonry sand), while ASTM C780 states the method of obtaining samples for compressive testing and the strength required for the mortar. Copies of these ASTM standards must be obtained to ensure full compliance with both the project specifications and industry standards.

7.4.3.5.4   Heating, Ventilation, and Air-Conditioning Testing
Although HVAC testing is performed by a professional testing agency, quality control personnel need to understand how and why these tests are performed so they can be performed at the appropriate times. The duct-work joint leakage test is performed after the ductwork is completed, prior to the insulation being installed on the outside of the ductwork.

7.4.3.5.5   Plumbing Tests
All pipes in the building must be checked for leaks. Testing for leaks involves subjecting all pressurized (supply/return/fire sprinkler) pipes to hydrostatic pressure testing, which is measured by a water pressure gauge. Usually, the test requires the pipes to hold 150 percent of the normal oper-ating pressure for two hours. Any drop in pressure indicates the presence of a leak in the line. Once this leak is found and repaired, the test is restarted for two hours. It should be noted that leaky joints must be tightened or taken apart and corrected. The application of pipe sealant to the outside of the pipe is not an approved correction method.

7.4.3.5.6   Performance Tests
Performance tests are required for many of the complicated systems that are installed in the building. A few of these systems include the fire alarm system, elevators, and water chillers/air-handlers. These types of tests are performed by the installer of the system and are only witnessed and veri-fied by QC personnel. Once again, it is important for QC personnel to have some sort of knowledge regarding what is involved with testing these sys-tems. The project specifications will state industry standards, which must be followed for proper testing.

7.4.3.6   When to Inspect Work

Knowing when to inspect works-in-progress is beneficial to the QC per-son. The following list is a summary of when and what to inspect on the jobsite.

7.4.3.6.1   Inspection Before the Commencement of Work
In some specific cases, as in the U.S. Army Corps of Engineers, this portion of inspection is called the "preparatory inspection phase." This inspection is made for each major work activity and is used to "verbally build" the item of work. A majority of the time, a preparatory inspection is held for each section. This involves holding a meeting to perform the pre-inspection of materials, methods, and personnel that are used to perform the work. Submittals and industry standards are used to verify that the work to be performed will be completed in compliance with the project documents. The use of sample panels for work such as masonry or stucco finishes is a prime example of this type of inspection. The workmanship and materials of the sample panel are inspected and approved prior to its implementation into the construction process. Corrections made at this level of inspection will cost less and will not impact the project schedule as much as if work was started before problems were discovered.

7.4.3.6.2   Inspection During Works-in-Progress
In some cases, the inspection of works-in-progress must be performed on a continual basis. QC personnel must maintain constant watch on work as it begins and heads toward completion. It is very important to verify that work starts out correctly, otherwise, reworks to correct the problem will occur. It is easier, and less expensive, to correct work as the work progresses instead of discovering defects after the work is completed. No one likes to perform the same item of work more than once.

7.4.3.6.3   Inspection of Work after Completion
Each work activity must also be inspected upon completion. This action is necessary to detect any deficient work prior to the next work activity to be performed. A "punchlist" consisting of the list of deficiencies discovered should be made and given to the parties responsible for the defective work. Verification that each deficiency has been corrected must be made to ensure that there are not any outstanding deficiencies. This stage of inspection will also require the performance testing of installed materials or equipment.

7.4.3.7   Paperwork/Documentation

Keeping track of quality control activities is an important duty of quality control personnel. Quality control paperwork is comprised of three types: recording logs, pre-installation inspection reports, and punch lists.

7.4.3.7.1   Recording Logs

Recording logs are used to keep track of items that either have been performed or are not completed as of yet. Submittal logs are used to keep track of the submittal flow throughout the course of construction. As each submittal is reviewed and a number is given to that submittal for tracking purposes. Depending on the submittal numbering system used, this submittal number will either be made up of a number or a number/letter combination. For example, suppose 20 submittals of information have been received from subcontractors/suppliers since construction began on a project. Then, the next submittal received will be labeled as "submittal #21." If this submittal is sent to the reviewing architect/engineer and comes back as «unapproved," this submittal must be resubmitted with the correct data. Then, this submittal will be labeled as "submittal #21A". A variation of the simple numbering system described above is used to keep track of submittals.

Each submittal that is received is listed on the submittal log under its given submittal number. Spaces for information regarding a description of the submittal, number of shop drawings (if applicable), submittal originator, and pertinent dates should also be provided on this log. A code column should be included, stating whether the submittal is "approved" or "unapproved."

The U.S. Army Corps of Engineers uses the following submittal action codes in the submittal review process:

A. Approved as noted
B. Approved, except as noted
C. Approved, except as noted; resubmission required
D. Will be returned by separate correspondence
E. Disapproved
F. Receipt acknowledged*
FX. Receipt acknowledged does not comply with contract requirements
G. Other (specify)

With the exception of government construction projects, submittal action codes vary between projects. Thus, one must research for the codes that are used in the submittal process.

Each construction deficiency discovered on the jobsite must be documented to ensure that the proper action is taken to correct the deficiency. Construction deficiency logs are used in conjunction with a notice of construction deficiency to track the identification and correction of defective construction work/materials.

The notice of construction deficiency states the details of the deficiency, while the deficiency log tracks each deficiency until the problem is corrected. Information included on these forms should give a description of the deficiency, responsible party, and a description of corrective action taken. The use of these forms will help ensure that the correction of each deficiency is not forgotten.

The concrete placement log is used to keep track of the date, time, location, amount, and type of concrete poured on the jobsite. A space for listing the concrete testing lab and the concrete cylinder set number is also provided. This cylinder sets a number that is useful due to trying to match the concrete compressive test results received from the testing laboratory with the date and location that the representative sample was taken.

7.4.3.7.2   Pre-Installation Inspection Reports

Pre-installation inspection report forms are helpful due to trying to schedule an inspection for works-in-place prior to being covered up by the next phase of work. These forms are signed once the stated portion of work is completed. The general contractor's quality control personnel perform their final inspection once everyone else has "signed off" on their portion of work. However, it should be noted that quality control inspections have to be performed on a continual basis while work is being performed. These pre-work installation forms are used for final inspection purposes, not for the initial inspection of the work.

7.4.3.7.3   Punch List Log

The last document to be discussed is the punch list log. This provides a general form that is used to keep track of deficiencies pointed out during project closeout. Blank copies of this form should be used on the jobsite to keep a handwritten list of punch list items. Then, the list can be sorted by the names of the responsible parties and their respective punch list items.

7.4.3.8   Quality Control Plans

Quality control should involve company executives as well as field personnel. Quality control plans provide the written "reference" document for the implementation of the quality control program.

This plan must explain the duties and activities of the quality control personnel as clearly and concisely as possible. The following writing suggestions should be used due to drafting such a plan.

The plan should be from the different departments involved with the quality control process. Also, it includes the field office personnel and the participation of the owner, engineer, subcontractors, and suppliers.

Preparation and implementation of the QC plan must be more than a "cosmetic" fix. The quality control program may look good on paper, but it can only serve its intended purpose by daily execution of the stated quality control procedures.

The plan must be easily understood by the person, who is going to implement the procedure listed in the manual. Items that should be included are organizational charts showing the chain of command, explanation of duties, lists of procedures, and examples of documents to be used.

The plan must be kept up-to-date by reflecting all changes required to maintain effective quality control on the job-site. This may include using suggestions from the employees responsible for QC duties.

The following guidelines have been established by the U.S. Army Corps of Engineers to be used when writing an organizational chart for the QC department:

- Lines of authority
- QC resources
- Adequately sized staff
- Qualifications of QC personnel
- List of QC personnel duties
- Clearly defined duties, responsibilities, and authorities
- Deficiency identification, documentation, and correction
- Letter to QC personnel giving full authority to act on all quality issues
- Letter stating responsibilities and authorities addressed to each member of the QC staff
- Procedures for submittal management
- Submittals must be approved by the prime contractor before review by owner's representative
- Log of required submittals, listing all required submittals showing scheduled dates that submittals are needed
- Control testing plan
- Testing laboratories and qualifications identified
- Listing of all tests required as stated by contract documents
- Testing frequencies listed
- Reporting procedures
- Quality control reporting procedure addressed

7.4.4   Quality Assurance

The following is an example of the importance of quality assurance. You decide to perform a sewage system project and about seven years ago a contractor company built you the same system in high quality in the plan time and cost. You are responsible for the decision without any influence from others.

Is it a good decision to go directly to this company or not? Why? (Please answer these questions before going to the next paragraphs.)

There are now different multinational companies worldwide in different industries and one of these industries is construction. So, every company and every one of us are both a customer and manufacturer or service provider sometimes. For instance, the contractor company does the service to the client and, at the same time, this contractor company is a client of the manufacturer for the plumbing equipment, HVAC, ceramic tiles, and other materials and equipment required to complete the project. At the same time, the factory that sells the ceramic tiles is also a client to the mechanical spare parts company to maintain their machines working.

So, any defect to any one of the systems will affect all. It is obvious that the quality system should apply to all the companies and organizations, assuming that everyone in the company has a good quality system.

Every company should build its own system to ensure that the product and service is based on specifications, requirements, and satisfaction.

When the quality assurance system is strong, it means that if anyone in the organization moved or retired the quality system is the same.

As an answer to the first question, if this company is a family company with a father and a son and the son becomes lazy and doesn't care, the project could be in trouble if his company is sharing in any of the project activity. On the other hand, if he has a real quality system, you can deal with him but you should also do an audit as we will explain later.

On the other side, for the multinational company, the chairman is usually sitting in a country far away from the project, so quality assurance will be in a document that can be reviewed by an external or internal audit. If there are complaints about the company from the owner, the system should record and solve these complaints.

The purpose of quality assurance is as follows:

- To make sure that the final product is in conformity with the specifications and that the employment is highly qualified and able to achieve a high quality of the product through the administrative system
- To ensure the application of the company's fixed characteristics among all sectors in the factory, regardless of the presence of the same people

- The benefits of the application of quality assurance systems can be summed up in that it gives the ability to produce a product identical with the required specifications and also reduce manufacturing cost because it will reduce waste or defective products. In particular, projects have a major impact because in these projects the time factor is very important and may be the main driver of the project.

For example, due to the construction of hotels, the provision of any day of the total time for the project will have a significant return on the owner. The same is true for oil projects. Therefore, when reducing or not rejecting any product, time is not wasted in removing what had been done or repaired or in negotiations between the team of the contractor and the owner and the supervisory, achieving savings of the total time for the project.

If the product is proper, no part is rejected from it and a strong and good relationship is achieved between the seller and the client by reducing the number of complaints from the client. That relationship is very clear between the contractor and consultant or architect of the owner. The contractor provides the service or the work required of him with the presence of complaints and few observations that do not have a strong impact on the whole project from the supervising engineer, indicating that the quality of the work of the contractor.

In the case of repeated complaints from the contractor that have a strong impact on the project, the problems may occur in the project and the result is that he would not be called again in similar jobs. This is dangerous in the world of markets as the reputation and quality of the final product with a good relationship between the parties has a powerful and direct influence in the reputation of the company which provides the final product.

The quality assurance system is the basic way for any factory, construction company, or owner to have the ability to enter both internal and external competition.

7.4.4.1   Quality Assurance in the ISO

ISO 9001 gave details on the requirements of quality management systems. When there is a contract between the parties and the manufacturer who wants to display its capabilities, the emphasis is on its potential in design and manufacturing with high quality.

These requirements investigate the compatibility between the stages of design and manufacturing in terms of quality.

The ISO 9001 shows a model for quality assurance during the design, development, production, and use of the product.

7.4.4.2 The Responsibility of the Contractor (Manufacturer)

The definition of manufacturer in ISO 9001 refers to the one who engages in the manufacturing or supplying of the product, as well as the one who is supplying the service required of it. In all references of study, control and quality assurance have been identified. In the case of construction projects, the company is the contractor for the establishment and implementation as well as the engineering office, which offers the service represented in the design and preliminary engineering drawings.

It is clear that the first responsibility during the process of TQM is the responsibility of the manufacturer. Whether that is the office-engineering consultant or construction company, they must make sure the plant and anything or everything that comes out through this foundation must match the specifications required by the owner or the party of the owner if the owner is represented in a company or institution.

The result became moving different products between countries and continents with free trade. Now there is talk that the world is smaller through international trade agreements and its attendant laws and mechanisms help the trade between countries, which now has an impact on the industry with its various types. So it rests with you to achieve the requirements of the open market, which has led to fierce competition among different companies in the field of construction. We find the presence and proliferation of offices, international consulting, or multinational contracting companies exist on the map of competition in the Arab world.

The competition between these companies stems primarily from the followers of quality assurance systems. In fact, the competition, conflict, and their interaction led to the presence of some different administrative regulations.

For comprehensive quality and emphasized quality control, work had been done to make these systems helpful for companies to deal between different countries.

The overall quality assurance systems depend, basically, on customer satisfaction achieved by adequate revenues, which help them get a good reputation in world markets and allow them to compete.

The knowledge of environmental requirements of the state may affect the quality of implementation and the return of the project since the designer designs piles on the basis that they are close to the residential community and the work on these piles might exceed the permissible limits of noise. Thus, it requires a change in design, such as the use of piles of discharge.

There are some basic steps that should be important to the company, which deliver the service to improve quality. The main steps include that the senior management level would be interested in the importance of control and quality assurance through comprehensive quality leadership.

The second step is that the management level would provide an atmosphere that helps in dealing with the rules of quality assurance easily and make sure that all employees are following instructions and steps of quality assurance. These reasons are often constraints faced by the administrative level and they are raised by engineers and junior staff.

Senior management should pay close attention to the training process by organizing training courses for all the employees of an organization on the quality assurance procedures and technical labor in particular.

7.4.4.3   Responsibility of the Owner

It is noted that many problems arise because of the bad quality of the final product or non-conformity of the project to the required specification, which is the fault of the owner or applicant because he or she may not have defined the desired product or specifications clearly. Therefore, the contractor must have all the required and completed data and this is the responsibility of the Consultant Office of the owner.

Figure 7.1 shows the relationship between the three parties: the owner, the contractor, and the consultant. Here, traffic is clear in the exchange of information and transactions, so if one party breaches the terms of quality he will affect the other parties.

Based on the specifications, the contractor or manufacturer shall determine the price and schedule based on the quality of the product itself and

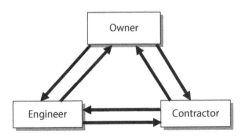

Figure 7.1 Relation between owner, contractor, and engineering company.

it is the responsibility of the owner to identify the required specifications of the project, strictly achieving its objectives.

The selection of the contractor or the manufacturer is one of the most important and most serious responsibilities of the owner or representatives of the owner.

Firstly, the owner starts the selection of the Engineering Office and then, at the start of execution, the owner has to choose the contractor, as these two selections are the most important factors and fundamental to the success of any project. Therefore, it is the responsibility of the owner or his representatives to gather enough information about the engineering office and the contractor and ensure from their previous work experience that they have performed the same project before. In the vital project, the owner can review the financial situation and make sure that the company is able to fulfill their obligations to the delivery of the project.

7.5 Project Quality Control in Various Stages

The project has been defined as a set of activities that has a beginning time and a time to be finished. These activities can be different from one project to another depending on the project. There are cultural projects or social projects, such as the literacy project, projects in engineering, such as the establishment of a residential building in an office, hospital, or the construction of a full apartment, or industrial installations, which are construction projects for roads, bridges, or railways. There are also irrigation projects and projects called civilian, but here we will focus on construction projects and, specifically, oil and gas projects.

Construction projects vary from one project to another depending on the size and value of the project. Therefore, the degree of quality control varies depending on the size of the project, especially in developing countries. Quality control may be sufficient in small businesses, but the contracting companies or small engineering offices that are aimed at the international competition are also increasing the quality of the projects, which increases the total cost of the project.

7.5.1 Feasibility Study Stage

Each stage of the project is important and has a different impact on the whole project, but differs depending on the nature and circumstances of the project and its value and purpose.

The feasibility study phase, followed by the stage of initial studies are those phases at the beginning of the idea that set the goal of the project and then select the ideas and engineering through the preliminary studies. The feasibility stage defines the goal of the project and the economic feasibility of any move. For those reasons, this stage requires full study, research, discussions, and deliberations with the economic data to compile the preliminary acceptance.

The feasibility study phase is followed by the emergence of the idea of the owner, and this owner may be an individual or a number of individuals. The study of the project will be performed by a competent consultancy office, as after these feasibility studies a strategic decision will be made. It is important to take into account that the consultant office has experience in the same type of projects where it depends on the expertise of the office in economic and follow-up the up-to-date factor that affects the economics of the domestic and global market.

The selection of the office that will do the feasibility study in the large-scale projects is important. Sometimes, you must hire foreign offices to carry out such studies, where the experience is important and vital.

7.5.2 Feed (Preliminary) Engineering

This is the second phase after the completion of the feasibility study for the project and goal setting. The phase of technical studies is no less serious than the first phase, but this phase of geometry is one of the most important and most dangerous stages of engineering. The success of the whole project depends on the engineering study. Therefore, this stage depends on the consultant office's specific experience in the type of project to be established. This study depends on the expertise of the engineering office, specifically in the project type where the experience of engineering projects varies depending on the kind of industrial project, whether it belongs to the petrochemical industries, oil and gas plant, power stations, or other industrial projects.

The required quality depends on the size of the project at this stage. For example, in the case of small projects such as building apartments, offices, or a small factory, the phase of initial studies is the ability of the office to determine the gender construction of the building. Will the project require reinforced concrete, concrete, precast or pre-stressed, as this determines the type of system of construction by the different types accepted, such as using solid slabs, flat slab, or hollow blocks, as well as the choice of system installation of the building such as columns and beams or the use of frames or shear walls.

These alternatives will vary by the size of the building itself and the requirements of the owner.

In the case of major projects, such as gathering people in a stadium or oil projects, the complexity of this phase increases with the interference of the studies of spatial principles. You can study landscapes and the different ways for the settlement of the land and the soil types on site, which determines the types of foundations that should be used for such trade-offs between the use of shallow or deep foundations, for example.

In the event of specialized projects, such as oil and gas projects, we need to study carefully the way to transfer the product and the differentiation among the alternatives in choosing the appropriate method.

From above, it is clear that as a result of the seriousness of that stage, experienced personnel is critical, as well as the owner having the capability to follow up on the initial studies in order to achieve the goal of the project and link the elements, such as the studies of civil, mechanical, electrical, and chemical engineering.

Generally, regardless of the size of the project, it must be based on the engineering preparation of the document to determine the requirements in the Statement of Requirement (SOR). The SOR contains the objective of the project and the needs of the owner and is an important document in the quality assurance system, as this document contains all the owner information.

The SOR is not only required for new projects or large-scale projects, but should also be used when making some modifications to the structure. In the case of a residential building, the owner should determine the number of floors required, the number of apartments on each floor, and the number of shops and any other useful requirements.

Upon receipt of the engineering office, the study is based on the SOR and is presented in a document called the Basis of Design (BOD). Through this document, the Engineering Consulting office should clarify the code and specifications that will work in the design, as well as the equations that will be used. It should also name the computer software that will be used, the required number of copies of drawings, and the sizes of the drawings. Additional information may be recalled, such as the information on weather, Met-ocean data, and survey data. This document will be reviewed in the same manner and, by this time, the owner will have determined the dates and number of revisions.

At this stage, it is important to make sure that both the owner and the engineering office have the same concept and there is full agreement between all the technical aspects.

Feed Engineering should be sent any prepared drawings so that they can be reviewed and carried out. Note the specified period of time agreed upon in advance, then return the drawings to the engineering office of the owner. The comments shall be returned to the owner until it reaches the final stage.

This phase may take months in the case of large projects and, therefore, the technical office and owner must have engineers who have experience in controlling costs and follow-up on time according to the schedule agreed upon in advance. The engineer is responsible for the cost, the estimated cost, and expected cost of the project and comparing them with what is in the feasibility study by the time we achieve the sequence stages. The estimate cost accuracy will be increasing with time until we reach the end of the preliminary studies, where one can obtain the nearest possible resolutions for the cost and time schedule.

It is imperative that at this stage you should not lose sight of determining the way to maintain the building in the future. This is done by determining the project lifetime. It must define the rules of construction, type of structure, and method of maintenance, as the site of the project itself and the surrounding environment must be protected from weather, which reduces the cost of maintenance over time. We may choose protection systems by using stainless steel, for example, through a system of protection high costs at the beginning of construction with a periodic maintenance, which is simple with few costs. On the contrary, we can use a protection system with low capital cost, but it will be taken into account in the annual budgets as the annual inspection and maintenance cost is high. In addition to that, the study takes into account the importance of structure itself, its geographical location, and ease of maintenance work and this philosophy also applies to the initial selection of mechanical equipment. For example, we may need a machine turbine with a high initial cost of an annual cost with a simple return. You may choose the frequency of a machine with a simple initial cost where the annual cost is high.

The importance of origin is the same and its mode of operation and maintenance affect the preliminary design. For example, in power stations and major projects you must ask whether it can fix the water tank or provide maintenance work and clean up. The answer to this question is a decision on whether an additional reservoir was needed.

If it's not needed, other design principles must be decided on at that stage. So, this stage requires very high experience since any error could result in future problems after operation, which could put the project at heavy losses that could have been easily avoided by simple solutions with a low cost from the beginning.

7.5.3 Detailed Engineering Study

It is assumed after the end of this stage that one can get the complete drawings and the full specifications for the whole project, containing all the details that make the contractor able to carry out the works. Those drawings are called the Construction Drawings.

Therefore, this phase requires many working hours, extensive contracts, good coordination, and excellent organization. A good manager allows freedom of communication between individuals and allows review with strong and continuous coordination.

It is noted that the complexity of this stage needs an affirmation of the quality.

Engineers always believe in utopias where our lives are dependent on the accuracy of the accounts and reviews, but the teamwork is not always so accurate. Often, work reaches someone late or you may have to take actions to correct it, or there may be some change in procedures within the company or department without your knowledge. All of this leads to a loss of time and, therefore, we believe that the overall atmosphere in which we work needs to be reformed and this itself is a vision of the system of quality assurance.

The system of quality assurance provides stable functioning of all departments, however, there may be a change in personnel. This problem often occurs at the stage of studies where you need it for intensive cooperation.

For example, when there is a strong relationship between the managers of the departments of civil and mechanical, you will see a sharing in information, the work will run smoothly, and there will be periodic meetings and productive correspondence.

The system of quality assurance at this stage is important because it is a process of organizational work. Everyone should know the goal of the foundation, the goal of the project by the institution, the responsibility of each individual, and the concept of quality is clear at all times and is supported by documentation.

Documents are considered the operational arm of the quality application process and must, therefore, be in the event of any amendment or correction in the drawings. When you set up the drawings, they must be received at a specific time for the owner to review, discussion must take place with any amendment, and remarks must be made to identify changes.

Through the development of a particular activity, the activity may be canceled, so there should be a quality system procedure to avoid confusion with the other copies of drawings to eliminate the human error. The modification revision number will continue to be updated and this system will continue until we reach the final stage of the project and take the final approval of the drawing with a sealed stamp that indicates that it is the final drawing for approval of construction. Approve for Construction drawings are obtained after the completion of the study phase. The start of the execution phase should have specifications and drawings fully ready to start the construction phase. You can imagine that some projects may have hundreds of drawings, special specifications, and other operation folder manuals, as well as for maintenance and repair in the event of some failures or trouble shooting.

Generally, it can be summarized in the following principles of design, which are divided into five aspects and should be covered in each quality assurance system:

1. Planning, design and development – Determine who does what in the design.
2. Entrance design – Be sure you know what the client wants in the design.
3. Troubleshooting design – Provide clarity for the final form of the design.
4. Verification of the design – Review with the client to make sure that the design is consistent with the needs of the client.
5. Change Design – Ensure that any change in design will be adopted by responsible persons.

7.5.3.1   Design Quality Control

The goal of quality control during the operational phase of studies in ISO 9001 (ISO 9001 Section 4.1) is to control the design at various stages, which has already been explained and is illustrated in the following form:

Requirements, design, development, planning, responsibilities, procedures and specifications, references, research marketing, and safety

Table 7.1 Design QC in ISO 9000.

Design Input	Output Design
• Instructions and control operations. • Taking into account the marketing materials • Specifications and tolerances allowed • Health and Safety and Environment Computer	• Design review • Design review process • Verification of the design

features, and safety and health overlap between the technical teams to change the design analysis of landslides in the internal audit.

The input design is all the technical information necessary for the design process, as it is clear that the foundation that is the information from the owner and SOR must be identified and developed to verify the accuracy of input design to ensure that the output of the process design is identical with the design required. Any information that is not clear or skeptical must be resolved at that stage.

Instructions to control operations are often stipulated in the contract where the client puts some instructions to control the whole process or request some specific actions.

Marketing is an important factor in the design process and must be compatible with the design and marketing research commensurate with the demands of the market. Therefore, we must make sure that the marketing team has entered at that stage.

The designer must take into account the available materials in the market and the abilities of the owner.

Specifications and tolerances must be identical with the design specifications of the project and the tolerances must be compatible with the specifications and requirements of the owner.

The designer must take the standards of health, safety, and the environment into consideration in design for the work at the site during the design phase.

Modification and renewal by the drawings have become easy with (CAD) Computer Aided Design in order to obtain more accurate information and access to information through various forms of tables and graphs.

Output design, a pre-final in the design process, must be a design compatible with all design requirements through internal audit. A change or modification in the design should not be compared with the old design because the design must be compatible with the specifications of safety and health.

Design review must be set in the early stages of design. Make sure that the design is done according to the constraints of the timetable and costs.

Verification of the design, in this case, is reviewed in design calculations for accuracy and is also done periodically, taking into account previous similar designs that can be created as a model for the design and operation test.

7.5.4 Execution Phase

Now everything is prepared for this stage. This stage requires both quality assurance and quality control. In the work of reinforced concrete structure, concrete itself consists of various materials such as cement, sand, gravel, water, and additives in addition to steel reinforcement and, therefore, must be controlled in terms of quality of each article separately as well as in the same mixture. The above must be done for quality control during the preparation of wooden forms and assembling the steel, casting and processing.

Thus, the contractor should have the administrative organization well organized in order to achieve quality control in addition to the existence of documents that would identify the time and date when work is carried out to determine the number of samples of concrete, which is to test pressure resistance and to identify the exact time, date, and test result.

Often, during construction there are some changes in the drawings as a result of the emergence of some problems at the site during the implementation or appearance of some ideas and suggestions that reduce the project time. Some changes do not negate the imbalance in the quality assurance system in the design stage.

After the work, the change must be made in the documents and post implementation is modified to get the drawings identical to the site of the as-built drawing.

The supervisory authority and the owner must both have their own organization, and the two most common cases in projects are the following:

1. Owner has the supervision team on site.
2. Owner chooses a consulting office that performs the design and handles the supervision.

In both cases, there should be strong organization similar to the organization of the contractor, as it is the tool that controls quality. But, when the contractor has a working group who has full knowledge of quality assurance and control, the dispute between the supervisory and the contractor will be narrower, where the controversy will not be about the final quality of the project. It is very important to the owner that the project is in good shape in the end.

You can imagine that if the owner has a competent staff and a strong knowledge about QC and QA, while the contractor staff has no knowledge about it, there will be more trouble that will affect the project as a whole.

The construction phase shows the strength of the contractor. For international contractors, the concept of quality assurance is quite clear because all the competitors on the international scene, for some time, have been working through an integrated system aimed at assuring the quality of work and set quality in all stages of implementation in order to achieve total customer satisfaction.

7.5.4.1   ISO and Control Work

The processes of execution have been developed (ISO 9001 Section 4.9) and are illustrated in implementation requirements for access to the required quality.

- Control the execution of special operations, environmental conditions, specifications, work instructions, procedures, control, and follow-up.
- Companies that operate through ISO should develop a plan for execution processes, structures, and construction that affect the final quality of the project.
- All the actions of the execution process are registered and have their own documents and you should make sure that there are adequate procedures to ensure that equipment is working properly for the duration of the project without any malfunction affecting the workflow.

7.5.4.2   Inspection Procedures

The phase of execution will be supplied by all required materials from many different locations in addition to the installation of these materials and it is the responsibility of the general contractor to make sure that the supply of various materials, as well as the construction and installation must be done through the required quality.

In the case of construction and installation, there must be specific instructions to determine the manner in which they work and the proper equipment to be used.

Investigation is the final quality process of the work or the quality of the materials supplied. It requires continuous inspection and testing work and the inspection must specify the following:

- The substance to be tested
- Test procedure
- Equipment required for testing and inspection and calibration of such equipment
- Inspection method
- Environmental conditions required, which will maintain it during operation, inspection, and testing

- The sampling method or the way to choose the appropriate sample
- Defining the limits of acceptance and rejection of the samples tested

The following items are from the ISO 9000 Section 4.11 and the inspection and measurement test.

- Control inspection
- Measurement and test equipment
- Calibration, maintenance, and the surrounding environment and storage and documents
- Registration and inspections

7.5.4.3   Importance of Contracts in Assuring the Project Quality

Note the diverse type of contracts between the owner and the contractor, as well as between owner and engineering office. A contract must be defined by the type of the project, which invokes the experience of the owner and the department. The existence of a bug in the contract will cause a lot of problems that may be difficult to resolve and the additional time will affect the final cost of the project.

Therefore, the number of contracts and the review of the included objective of the project needs to be highly experienced.

In general, the contract documents contain the drawings and specifications for the materials, labor, and tools used and must identify the conditions of employment and position, in which the relationship between the owner, contractor, supervisory, and engineering facility is defined in the decade list, along with the quantities of each item and price, as in the case of indexation.

Other essential items are usually overlooked by the engineers in oil and gas projects, such as the requirements for health, safety, and the environment, which determine the disposition and the cost in the case of injury or sickness at the site and calculate taxes and courts that are referenced in the event of disagreement, etc. Often overlooked are administrative items that are not as important as the engineering items that have been mentioned.

7.5.4.4   Checklists

The ISO 9001 selects who conducts the review process and some of the menus that contain questions from the manufacturer or internal departments. When you read the questions in the lists, you will find that they cover many important aspects of basics and quality control in all stages of the manufacturing or implementation of products.

Checklists contain the following specific questions, which are with the auditors of public review of the company's performance with other lists in the special design phase and implementation phase and those lists are detailed with questions to control the full and comprehensive review of the quality system at the stage of design and implementation.

7.5.4.4.1   Checklists for Reviewers

This is an itemized description of the item questions.

1. Management Tasks
- Is this representative of the department responsible for achieving ISO 9000?
- Is it the specific responsibility and authority of all influential individuals in the quality?
- Are the available resources adequate?
- Has skilled labor been engaged in the work related to their skills?
- Is the audit done by independent auditors?

2. Quality Management System
- Is the establishment of a system and the number of complete documentation of the quality management system to make sure that the product will achieve specifications?
- Is there an audit of the contract for the review of contracts and to achieve cooperation between the activities to implement the contract?

3. Control Design
- Are individuals eligible to complete the work assigned to them?
- Are the design requirements reviewed by the owner?
- Is the output compatible with the design input?

4. Control of Documents
- Is there a procedure to control all documents' quality?
- Have the documents been reviewed and approved before use?

5. Purchases
- Do you buy products in conformity with the requirements of quality?

6. Supplier of the Product Supplies
- Is there a procedure to verify the storage and maintenance of products supplied by the supplier?

7. Control in Manufacturing or Implementation
- Is the manufacturing process or implementation specific and planned through the documents with instructions to work?
- Do the procedures control manufacturing and implementation of special operations?

8. Inspection and Testing
- Whether a procedure is to ensure that the product has not been used before, testing or verification that they were identical with the specifications is necessary?
- Is there a procedure to verify conformity with the product specifications during manufacturing?
- Is there a procedure for final inspection tests?

9. Test Devices
- Is there a procedure to control the calibration and maintenance of tests?

10. Test Results
- Is there a procedure to make sure that the product non-conformity with the specifications is controlled?

11. Step Corrective
- Is there a procedure during inspection to determine the cause of the incompatibility of the product with customer requirements and specifications and to determine the steps necessary to fix it and avoid the problem occurring in the future?

12. Entrepreneurship and Storage
- Are there procedures and documentation for the control of contractors, storage, and handling?

13. Quality Recording
- Is there a procedure to maintain the action and to identify, collect, and store all documents related to quality?

14. Review of Internal Quality
- Is there a planned system of internal audit documents?

15. Training
- Are there training procedures for identifying training needs for employees to have an impact on quality?

16. The Service
- Are the service procedures present to achieve service?

17. Statistical Processes
- Are there procedures to determine the statistical information required to accept or reject the product?

7.5.4.4.2   External Auditing

A team outside the company usually performs external auditing. A company that takes the ISO should also be a company whose products and safety we can trust. But, in reality, in some countries you cannot have this list and in other countries the companies have an ISO, but they don't apply any of the system for quality assurance, as most of the companies are focused on quality control rather than the quality assurance or developing a quality manual.

Therefore, in major projects the owner should formulate the team from his quality department to make external auditing for any company, engineering, contractor or service provider.

To be sure that the company will be well prepared for this visit from the owner auditing, he or she will gain a lot from the big project, but it will be difficult in small projects. So, to control this issue a company needs to be registered with your company and they should provide their prequalification. At this time, you can perform the auditing to decide if they will be registered with company's vendors list or not.

The team from the owner company should be competent and have strong skills and knowledge in the quality management system and how to audit the other company in their country or overseas countries.

First, the contractor or the service provider will deliver his quality manual, which will be reviewed by the quality team. Then, they will visit the site with a representative from the company to show their system in action and to inspect it. The process is as follows:

- Visit the supplier site to perform a complete inspection.
- The supervisor will describe to the team exactly how their Q.C. system works.
- The contractor provides examples of Q.C. documentation.
- It is possible for the team to ask for a previously inspected batch to be rechecked.

- Check if the test equipment is regularly maintained.
- The rejected or unacceptable products are clearly marked and segregated to avoid any chance of their accidental inclusion with acceptable products.

Table 7.2 Design check list.

Item	Questions	Yes/No	Remarks
1	Do they have a system to assure the client presents his or her needs clearly?		
2	Are the client requirements clear to all the design members?		
3	What is the international standard and specification they use?		
4	Are these standards available in their office?		
5	Are the drawings and documents sent by the client registered?		
6	Do they have a document management system?		
7	Do they define the name of the discipline lead?		
8	Are the activities clear to them?		
9	How can they select the new engineers?		
10	Do the drawings have a number?		
11	Are there strong numbering systems to the drawings?		
12	Do they prepare a list of the drawings?		
13	Are they updating this list?		
14	What is the checking system in calculation?		
15	What is the checking system in the drawings?		
16	Are the employees familiar with CAD?		
17	Do they have a backup system to the documents and drawings?		
18	Do they have antivirus?		
19	Do they use a sub engineering office?		
20	What is the method and criteria for the selection?		
21	Is there a good relation between the design team and the supervision team on site?		
22	Are they experienced with the technical inquiry?		

After the site visit, there are three probable outcomes:

1. Acceptable to be registered in company bidder list
If the evaluation has shown that the supplier has a satisfactory Q.M.S., there are no deficiencies and the supplier is able to give an assurance of quality.

Also, in this case the supplier may have proven that they are up to satisfactory standards.

2. Weak Quality System
If the team finds several significant weaknesses in the supplier's system, the supplier will have to take steps to overcome these failures and improve their Q.M.S.

Table 7.3 Construction checklist.

Item	Questions	Yes/No	Remarks
1	Is there a quality control procedure?	1	
2	IS the QC procedure understood by all members and how?	2	
3	Does the QC match with the task?	3	
4	What is the way to assure with the tests?	4	
5	Is there equipment to do the test calibrated?	5	
6	How they do the test piping, concrete, welding, and what is the confidence for this test?	6	
7	Is a third party used for the test?	7	
8	What is the criteria for choosing the third party?	8	
9	Do they use a sub contractor?	9	
10	What is the criteria for choosing these sub contractors?	10	
11	Do they regularly maintain their equipment?	11	
12	Do they have a certificate for the cranes and wire?	12	
13	Is there a team on site that knows if the project is time or cost driven?	13	
14	Does their team know the project objective and target?	14	
15	Do they control the documents and drawings on site?	15	
16	Do they have a document management system?	16	

The supplier can ask for another evaluation to confirm that their quality is approved. To enhance their company and employee work will take time, so it is not preferred to register the company but to give it a period to change.

3.   Unacceptable Quality System
In this case the team found that the supplier will have to make radical changes to improve their overall Q.M.S.

Note that in this case the actions from the supplier to reach the target to satisfy Q.M.S. will take not less than one year, so avoid dealing with this supplier.

ISO 9001 provides a check list that is essential and important for the auditing team. The following is the check list for every phase of the project. In the design phase there is a sample for what will be asked from the engineering company. For the construction phase, this check list is tailored to be a sample when auditing a contractor company.

From the above, it is important that all employees in the project have very good technical skills plus the quality system, as is stated in ISO 9001 section 4.18. The should be required in training to increase their technical and managerial skills for quality control in addition to all the knowledge about total quality management.

7.6   Operational Phase of the Project

The owner is responsible for the operation phase, as in this phase the owner will have full authority and responsibility to operate the project after the commissioning and start-up phase.

During operation there is usually a requirement to modify the facilities due to operation needs.

This is traditionally in the oil and gas industry in the case of an increase in production or a need to modify the mode of operation.

In some regular cases, workshops are extended in normal industry. Most international companies that follow ISO have a management of change system and procedure.

In this procedure, the required modification is identified and approval is granted from all the engineering disciplines. In new projects, it is preferred to go to the original engineering office that performed the design to do the engineering for this modification.

An example of bad management of change (MOC) in an international hotel is the conversion of one room from normal use to be a planet land for entertainment. They put clay on the floor without performing any management of change process and, due to this heavy load, the floor collapsed and damaged four cars in the garage underneath the room.

From previous experience, projects in every step of engineering and construction concentrate on these goals and, usually, the input from the operation is very little.

In most oil and gas projects, the owner creates a separate organization for this project. In most cases, the project is under the responsibility of different departments. However, the operation is the end user, but usually a limited number of operation engineers share in the project for many reasons, as the operation cannot release. On the other hand, the operation members who are usually one or two engineers just define what they need. In this circumstance, there is usually an expectation for not full satisfaction from the operations department.

7.7   Total Building Commissioning System

This system is very important in oil and gas projects. However, those in the oil field are more familiar in commissioning and there are many professionals in this area, but they apply in the start-up of a project to be a link between the projects staff and operations staff. In fact, the USA applies this system for normal buildings to formulate the team responsible for commissioning from the engineering phase until the operation phase.

Per our discussion in Section 7.6, in some cases the operation is not satisfied with the project output. The optimum solution is to apply a "total building commissioning system," which is already applied in the USA for residential, administration, commercial, and public buildings.

Historically, the term "commissioning" has referred to the process by which the heating, ventilation and air conditioning (HVAC) systems of a building were tested and balanced according to established standards prior to acceptance by the building owner.

Today's use of commissioning recognizes the integrated nature of all building systems' performance, which impacts sustainability, workplace productivity, occupant safety, and security.

In the USA, the U.S. General Services Administration (GSA), through its Public Buildings Service (PBS), manages buildings that house over a million federal associates and has an on-going planning, design, and construction program to meet the federal customers' housing needs.

PBS' project delivery program is the vehicle for transforming our customer agencies' vision into reality. The built environment for the nation's public buildings, including courthouses, federal office buildings, laboratories, and border stations, in turn shapes the communities and landscape in which they reside.

In our case the engineering team in the owner company will have the same responsibility as GSA.

The National Conference on Building Commissioning has established an official definition of 'Total Building Commissioning' as follows:

> "Systematic process of assuring by verification and documentation, from the design phase to a minimum of one year after construction, that all facility systems perform interactively in accordance with the design documentation and intent, and in accordance with the owner's operational needs, including preparation of operation personnel."

What we need here is to have a wide definition of "total facilities commissioning," which would refer to any discipline activity sharing in the project.

With my experience in the major projects in the engineering phase, the engineering company worries about man-hours and approves the bill and all the engineering disciplines focus on the design matching with the standard and specification demanded by the owner. On the other hand, the owner's representative engineering staff focuses on the deliverable of the engineering, which is strong, technically, and matches with the standard and specification. In reality, it will be a representation of operations in the engineering group, but unfortunately one or four is not enough and with the time that will be focused on technical issues, the only objective will switch to being the engineer within standards and to finish the review cycle on the time as the direct manager is focused on cost and time. The project needs a separate entity always looking to the owner requirements.

Below is information from the GSA document, tailored and changed to be match our needs in oil and gas and other industrial projects.

- Philosophy provides GSA's definition and expectations for commissioning.
- Building Commissioning Process details the considerations, practices, and recommendations for commissioning along the GSA project process including Planning, Design, Construction, and Post-Construction Stages.
- Appendices provide samples, tools, definitions, and links to further resources for additional information on commissioning.

Benefits of Commissioning for GSA Buildings
- Improved facilities, productivity, and reliability
- Lower utility bills through energy savings
- Increased operations and owner satisfaction
- Enhanced environmental, health conditions, and occupant comfort
- Improved system and equipment function
- Improved facilities operation and maintenance
- Increased operations safety
- Better facilities documentation
- Shortened transition period from project to operation
- Significant extension of equipment/systems life cycle
- GSA defines the benefits in industry sources and indicates that, on average, the operating costs of a commissioned building range from eight percent to twenty percent below that of a non-commissioned building. This philosophy applies for the plant buildings only.

7.7.1 Planning Stage

Figure (7.2) presents the main steps in applying a building commissioning system in planning, design, construction, and post construction phases. These steps will be discussed in detail in the following sections.

7.7.1.1 Identify Commissioning Team

The first step in the commissioning process is for the GSA project manager (GSA PM) is to identify and layout the makeup of the Commissioning Team. The exact size and members of the commissioning team will vary depending on project type, size, and complexity.

In general, the team will consist of the following:

- GSA Project Manager (Team Leader)
- GSA Operating Personnel
- Customer Agency Representative(s)
- GSA Technical Experts (i.e., Structural, Mechanical, Electrical, Fire Protection, Elevator, Seismic, LEED/Sustainability, etc.)
- Construction Manager (CM)
- Construction Contractor and Subcontractors
- Commissioning Agent (CxA)
- Architect/Engineer (A/E)

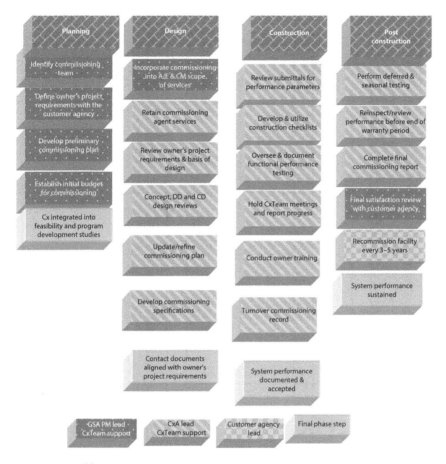

Figure 7.2 Building commissioning system main steps.

The Responsibility Definition is presented in Table 7.4. The following definitions apply to the Roles and Responsibilities Matrix:

- Lead (L) = Direct and take overall responsibility for accomplishment
- Support (S) = Provide assistance
- Approve (A) = Formally accept either written or verbal, depending on the situation
- Participate (P) = Take part in the activity (i.e., attend the meeting, etc.)

Table 7.4. Commissioning roles and responsibilities matrix.

Legend L = Lead P = Participate S = Support I = Inform A = Approve V = Verify	GSA Project Manager	GSA Operating Personnel	Customer Agency Reps	GSA Technical Experts	CM	Construction Contractor	CxA	A/E
Planning Stage								
Identify Commissioning Team	L/A	S	S	P/S				
Develop Owner's Project Requirements	L/A	S	S	S				
Develop preliminary commissioning scope	L	S	S	P/S				
Develop Preliminary Commissioning Plan	L	S	S	S				
Establish budget for all Cx work & integrate costs for commissioning into project budget	L	S	S	S				
Include time for Cx in initial project schedule	L	I	I	I				
Include Cx responsibilities in A/E & CM scope of services	L/A	S		S				
Design Stage								
Contract for Commissioning Agent Service	L/A	P	P	P	L			
Hold Design Stage Cx meeting	P	P	P	P	P		L	P
Identify project specific responsibilitie	L	P	P	S	S		P	P
Review Owner's Project Requirements documentation for completeness & clarit	S	S	I		I		L	I
Develop Basis of Design	A	P	P	S/A	I		I	L
Perform focused Cx reviews of design drawings & specification	P	P	P	P	S		L	S

Activity								
Perform project constructability review	P						I/S	S
Incorporate appropriate changes to contract documents based upon design review	A	P	P	I	I		L	L
Refine Owner's Project Requirements based upon Design Stage Decision	A	P	P	S	I		L	S
Create Cx specifications including testing protocols for all commissioned equip./system	I	I	I	P/S	S		L	S
Integrate Cx activities into project schedul	A	I	I					
Coordinate integration issues & responsibilities between equipment, systems & discipline	A	I	I	P/S	S		V	L
Update Commissioning PlaN	A	I	I	I	S		L	I
Incorporate commissioning requirements into Construction Contractor's Scope of WorK	A			I	L		S	S
Construction Stage								
Revise Commissioning Plan as necessary								
Review submittals applicable to equipment/systems being commissioned								
Review project submittals for construction quality control & specification conformance	A	I		I	I	S	I	L
Develop functional test procedures and documentation formats for all commissioned equipment & assemblies	I			P	A	S	S	L
Include Cx requirements and activities in each purchase order and subcontract written	I			I/P	A	L	S	V

(Continued)

Table 7.4 (Cont.)

Legend L = Lead P = Participate S = Support I = Inform A = Approve V = Verify	GSA Project Manager	GSA Operating Personnel	Customer Agency Reps	GSA Technical Experts	CM	Construction Contractor	CxA	A/E
Develop construction checklists for equipment/systems to be commissioned	A			P	I	I	I	L
Install components & system	I	I	I		A	A	L	V
Review RFIs and changes for impacts on Cx	A	I	I	I/S	S	L	S	V
Demonstrate operation of systems	I		P/I		I	P	L	V
Complete construction checklists as the work is accomplished	I	I		I	I	S	L	A
Continuously maintain the record drawings and submit as detailed in the contract documents	A	S			I	S	L	V
Coordinate functional testing for all commissioned systems & assemblies	I	I	I	P/A	I	S	S	L/A
Perform quality control inspections	I	I	I	I/P	I	L	S	P/I
Maintain record of functional testing	I	I	I	I/P	I	S	S	L
Prepare Cx Progress Report	A	I	I	I/P	I	P	S	L
Hold Construction Phase Cx meeting	P	P	P	P	P	P	P	L
Maintain master Issues Lo	I	I	I	I	I	S	I	L
Review equipment warranties to ensure GSA responsibilities are clearly define	I	I				S	S	L
Implement training program for GSA Operating Personnel	I	P	P	I/S	P	S	S	L

Compile and deliver Turnover Package	A	A				S	S	
Deliver Commissioning Record	A	P	P	I	S	S	S	L
Post-Construction Stage								
Coordinate & supervise deficiency correction	A	P		I	I/S	L	S	I
Coordinate & supervise deferred & seasonal testing	A	P		I	S	S		I
Review & address outstanding issue	A	P		I	I/S	S	S	I
Review current building operation at 10 months into 12 month warranty period	A	P		I	S	S		I
Address concerns with operating facility as intended	A	P		I	S	S	S	S
Complete Final Commissioning Report	A	P			I/P	I		I
Perform Final Satisfaction Review with Customer Agency 12 months after occupancy	A	S	S	S	S	S	S	S
Recommission the facility at 3–5 years after turnover to reset optimal performance	L	P		L	P			S

L = Lead P = Participate S = Support I = Inform A = Approve V = Verify

CM = Construction manager

A/E = Architectural and engineering

CxA = commissioning agency

- Inform (I) = Make this party aware of the activity or result or provide a copy of the deliverable
- Verify (V) = Confirm the accuracy or completeness of the task

7.7.1.2   Define Owner's Project Requirements with the Customer Agency

The objective of commissioning is to provide documented confirmation that a facility fulfills the functional and performance requirements of GSA, occupants, and operators. To attain this goal, it is necessary to establish and document.

The owner sets project requirements and criteria for system function, performance, and maintainability. The Owner's Project Requirements will form the basis from which all design, construction, acceptance, and operational decisions are made. The following suggested categories provide a framework for the types of requirements that shall be considered.

7.7.1.3   Develop Preliminary Commissioning Plan

The Commissioning Plan establishes the framework for how commissioning will be handled and managed on a given project. This includes a discussion of the commissioning process, schedule, team, team member responsibilities, communication structures, and a general description of the systems to be commissioned. This preliminary version of the plan shall be developed by the GSA PM in conjunction with the Customer Agency. The suggested structure of the Commissioning Plan is as follows. All information in the Commissioning Plan must be project specific.

7.7.1.4   Commissioning for Certifications (LEED, Energy Star, etc.)

For a housing building, the building commissioning system looks for the building design that will achieve energy saving through leadership in Energy and Environmental Design (LEED). Any building will be certified as a green building if its design achieves points to be a gold, silver, or platinum building and some of its points are gained by using the total building commissioning system.

Development of the preliminary Commissioning Plan and initial commissioning scope shall also include a discussion regarding project certifications and goal attainment (i.e. LEED, Energy Star, Energy Goals, Design Awards, etc.).

Table 7.5 Owner requirements.

Accessibility	Access and use by children, aged, and disabled persons
Acoustics	Control of internal and external noise and intelligibility of sound
Comfort	Identify and document those comfort problems that have caused complaints in the past and which will be avoided in this facility (i.e., glare, uneven air distribution, etc.)
Communications	Capacity to provide inter- and intra-telecommunications throughout the facility Constructability Transportation
Constructability	Transportation to site, erection of facility, and health and safety during construction
Design Excellence	Potential/Objectives for design recognition
Durability	Retention of performance over required service life
Energy	Goals for energy efficiency (to the extent they are not called out in the Green Building Concepts)
Fire Protection and Life Safety	Fire protection and life safety systems
Flexibility	For future facility changes and expansions
Green Building Concepts	Sustainability concepts including LEED certification goals
Health & Hygiene	Protection from contamination from waste water, garbage and other wastes, emissions and toxic materials
Indoor Environment	Including hydrothermal, air temperature, humidity, condensation, indoor air quality and weather resistance
Life Safety	Fire protection and life safety systems
Light	Including natural and artificial (i.e., electric, solar, etc.) illumination
Maintenance Requirements	Varied level of knowledge of maintenance staff and the expected complexity of the proposed systems
Security	Protection against intrusion (physical, thermal, sound, etc.) and vandalism and chemical/biological/radiological threats
Standards Integration	Integration of approved federal, state, and local as well as GSA and customer Agency standards and requirements
Structural Safety	Resistance to static and dynamic forces, impact and progressive collapse

Table 7.6 Commissioning plan.

Purpose and general summary of the Plan	Introduction
Overview of the project, emphasizing key project information and delivery method characteristics	General Project Information
The commissioning scope including which building assemblies, systems, subsystems and equipment will be commissioned on this project	Commissioning Scope
Project specific commissioning team members and contact information	Team Contacts
Documentation of the communication channels to be used throughout the project	Communication Plan and Protocols
Detailed description of the project specific tasks to be accomplished during the Planning, Design, Construction and Tenant Occupancy Stages with associated roles and responsibilities	Commissioning Process
List of commissioning documents required to identify expectations, track conditions and decisions, and validate/certify performance	Commissioning Documentation
Specific sequences of events and relative timeframes, dates and durations	Commissioning Schedule

Table 7.7 Work during the commissioning process.

Appendices	Work completed during the commissioning process
A	Owner's Project Requirements
B	Basis of Design
C	Commissioning Specifications
D	Design Review
E	Submittal Review
F	Issues Log
G	Construction Checklists
H	Site Visit and Commissioning Meeting Minutes
I	O&M Manual Review
J	Training
K	Functional Performance Tests & Seasonal Testing
L	Warranty Review
M	Test Data Reports

For the project to be LEED certified, commissioning process activities must comply with the prerequisite requirements for fundamental building commissioning and the project team may opt to pursue an added LEED point for additional commissioning.

The process provided in this guide provides the necessary steps to comply with both prerequisite and additional commissioning requirements.

7.7.1.4.1   Establish Initial Budget for Commissioning

- Based upon the Preliminary Commissioning Plan, the GSA owner Project Manager includes budgetary costs for commissioning in the Feasibility Study and the Program Development Study. It is critical that the overall established budget, which is submitted for funding approval contains necessary monies for commissioning.
- Specifically, the Feasibility Study and Program Development Study deliverables, per the owner's Project Planning Guide, call for estimated construction costs (ECC) and estimated total project costs (ETPC). These estimates must include line items for both commissioning services and testing.

7.7.1.5 Commissioning Agent Costs

The following can be taken as a guide from GSA:

- Total building commissioning costs for (CxA) services can range from 0.5 percent to 1.5 percent of total construction costs (according to U.S. Department of Energy's Rebuild America Program, written by the Portland Energy Conservation, Inc. (PECI)).
- The National Association of State Facilities Administrators (NASFA) recommends budgeting 1.25 to 2.25 percent of the total construction costs for total building (CxA) services.
- GSA's commissioning practice is expected to cost approximately 0.5 percent of the construction budget for federal buildings and border stations.
- More complex projects, such as courthouses, could run 0.8 to one percent of the construction budget and even more complex facilities, such as laboratories, can exceed one percent.
- Factors influencing commissioning costs include facility type, phasing 24/7 operations, the level of commissioning desired, and the systems and assemblies chosen to be commissioned.

The above costs only cover Commissioning Agent fees. There are also costs to the Construction Manager, Construction Contractor, engineering office, and owner staff for sharing part in the commissioning process. The profile of these costs will vary depending on roles and responsibilities chosen. For a detailed estimate of professional service fees, an itemized level of effort needs to be performed based on unique project requirements.

The relationship between commissioning cost per square feet and the floor area is presented in Figure (7.3).

7.7.1.6   Cost-Benefit Analysis for Commissioning

Recent PECI studies indicate that, on average, the operating costs of a commissioned building range from eight to twenty percent below that of a non-commissioned building.

Discussing the cost data for office buildings suggests that building commissioning can result in energy savings of 20 to 50 percent and maintenance savings of 15 to 35 percent.

Beyond operating efficiency, successful building commissioning has been linked to reduced occupant complaints and increased occupant productivity.

7.7.2   Design Stage

Design Stage commissioning activities serve to assure that the owner's project requirements for items such as energy efficiency, sustainability, indoor environmental quality, fire protection and life safety, etc. are sufficiently defined and adequately and accurately reflected in the contract documents.

The Design Stage is the commissioning team's opportunity to assure that building systems and assemblies as designed will function according to user expectations.

Further, specific tests and procedures designed to verify the performance of systems and assemblies are developed and incorporated into the contract documents.

7.7.2.1   Incorporate Commissioning into A/E and CM Scope of Services

GSA commissioning activities may be more rigorous than A/Es and CMs typically included in their scope of services.

Design, construction, and post-construction commissioning activities must be defined and written into the architect/engineer and construction manager scopes of work and executed contracts.

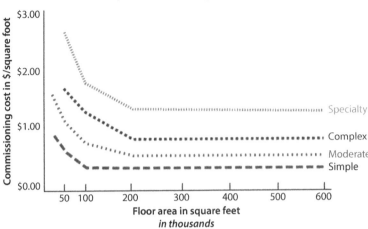

Estimates of construction phase commissioning costs

Figure 7.3 Relationship between the commissioning cost and the building area.

By this stage of project development, the GSA project manager must have an awareness of how commissioning services will be delivered. GSA's preferred method for engaging a commissioning agent is to arrange for the construction manager to contract directly with a commissioning agent. However, there will be exceptions based on project specific drivers. Should the project team determine that the CM will contract the commissioning agent, this must be written into the CM's scope of work.

Commissioning services for design and construction management professionals shall minimally include, but are not limited to, the items listed on the following page.

7.7.2.1.1   Design Professional

1. Participate and aid in the documentation of the owner's project requirements.
2. Document revisions to owner's project requirements and obtain GSA approval.
3. Document the basis of design.
4. Integrate Cx process requirements and activities provided by the CxA into the contract documents.
5. Attend commissioning team meetings (three design review meetings and monthly construction stage Cx team meetings).
6. Specify and verify that the operation and maintenance of the systems and assemblies have been adequately detailed in the construction documents.

7. Review and incorporate, as appropriate, the CxA's comments into the contract documents.
8. Participate in the operations and maintenance personnel training as specified in the training program.
9. Review test procedures submitted by the contractor.
10. Review and comment on the CxA's progress reports and issue logs.
11. Witness the functional testing of all commissioned systems and assemblies.
12. Review and accept record documents as required by the contract documents.
13. Review and comment on the final commissioning record.
14. Recommend final acceptance of the systems to GSA.
15. Verify systems are installed as specified

7.7.2.1.2 Construction Manager

1. If appropriate, lead the RFQ process for commissioning services and award a contract to a Commissioning Agent directly under the Construction Manager.
2. Include commissioning process activities and requirements into all general contractor bid packages.
3. Work with the commissioning team to develop a schedule for commissioning activities and incorporate commissioning activities into the overall project schedule.
4. Provide personnel with the means and authority to coordinate implementation of the commissioning process as detailed in the contract documents.
5. Attend commissioning team meetings (three design review meetings and monthly construction stage Cx team meetings).
6. Coordinate with the Commissioning Agent in the development of a commissioning plan.
7. Perform quality control functions, particularly in the areas of design reviews for constructability and inspection.
8. Participate in and assist with the functional testing of all commissioned systems and assemblies.
9. Provide technical expertise such as testing, cost estimating, and resolving disputes.
10. Coordinate and document owner/operator training.
11. Issue a statement that certifies all work has been completed and the facility is operational, in accordance with the contract documents.

12. Coordinate General Contractor remedies for deficiencies identified by the Commissioning Agent during their verification of the installation or tests.
13. Review and comment on the final Commissioning Record.

7.7.2.1.3   Retain Commissioning Agent Services

GSA's suggested practice is to have the CM hire a subcontractor to act as the Commissioning Agent, resulting in no additional contract management responsibilities for GSA.

In this case, the CM will lead the RFQ process for commissioning services. There will be exceptions to this suggested practice and, in these cases, GSA will lead the RFQ process for a Commissioning Agent. Regardless of the contracting method, the Commissioning Agent shall be on board by the beginning of Design Development.

7.7.2.1.4   Commissioning Agent Qualifications

The Commissioning Agent and the CM generally have different skills. In general, the CM provides management, technical, and administrative expertise during the design and construction phases to ensure that the customer agency's goals relating to schedule, budget, scope, and quality are met.

A CxA has technical background and in depth expertise with the commissioning process including verification techniques, functional performance testing, system equipment, and operation and maintenance (O&M) knowledge.

Furthermore, the CxA shall bring a total building commissioning perspective to the project, be knowledgeable in national building fire codes, as well as water-based extinguishing systems, detection systems, LEED, energy efficiency imperatives, and demonstrate experience with federal requirements (i.e., blast, progressive collapse, security, etc.).

7.7.2.1.5   Request for Qualifications (RFQ) for Commissioning Agent

The RFQ for CxA services is based upon the Preliminary Commissioning Plan and the commissioning budget established in the Program Development Study (PDS).

Depending on the CxA delivery method (i.e., CM versus GSA contracts), this may be the responsibility of either GSA or the CM.

7.7.2.1.6   Commissioning Agent Selection

This service shall be acquired in the same manner as other professional services. The Commissioning Agent shall be chosen on the primary basis of qualifications and not solely based on price. The involved work order and selection procedures should adhere to the involved provisions for work

Table 7.8 CxA selection.

Building type, square footage, general program, overall project Budget, milestone schedule dates, LEED and other certification pursuits, etc.	Project Background
GSA project objectives for commissioning	Objectives
Design, Construction, and Post-Construction Stage expectations for the Commissioning Agent	Scope of Work
Preliminary identification of the systems and assemblies to be commissioned. Once contracted, the CxA will further develop this matrix.	Systems and Assemblies
Desired qualification of the CxA	Qualifications
Expectations for format and content of prospective CxA's proposal	Proposal
Statement on GSA review of CxA changes in personnel for the project	Change in Personnel
A table indicating the selection criteria and scoring system for evaluating CxA proposals	Selection Criteria
	Proprietary Information (disclaimer)
	Protection and Control of Government Documents (disclaimer)

order issuance and be fitted with requirements/contents that are aligned with standard work order formats. National contracts are in place that can support commissioning services.

It is recommended that the CxA is contracted according to a two phase fee negotiation process.

The first phase includes Design Stage responsibilities and the second phase includes construction and post-construction activities.

The negotiation of the construction and post-construction stage fee is based upon a substantially completed design and the actual type and number of equipment, systems, and assemblies to be inspected, started, and tested.

Within the Design Stage proposal, the CxA shall be asked to provide budgetary numbers for the construction and post-construction stages.

7.7.2.1.7 Review Owner's Project Requirements and Basis of Design

- As described in previous sections, the owner's project requirements are developed as part of GSA's project planning processes and established baseline criteria for facility function, performance, and maintainability.
- The Basis of Design (BOD) is developed by the engineering office early in the Design Stage based on the owner's project requirements. It is the primary document that translates the GSA's and customer agency's needs into building components such as HVAC systems, the building envelope, security systems, building automation system, etc.
- The BOD describes the technical approach planned for the project, as well as the design parameters to be used. The BOD is typically developed by the engineering office and done in technical terms, whereas the owner's project requirements are developed by GSA in concert with the customer agency and expressed in layman's terms.

7.7.2.1.8 Review Owner's Project Requirements and Basis of Design

- Through the design process, a key role for the Commissioning Agent is to facilitate a clear understanding of expectations by the design team. To do this, the practice of conducting program review workshops is to be used to offer all stakeholders the opportunity to indicate what they want to see in the next design submission.
- The Project Planning Tools' Commissioning Tool identifies such practices in the work breakdown structure associated with defining roles and responsibilities.

7.7.2.1.9 Concept, DD, and CD Design Reviews

- The Commissioning Agent provides three focused reviews of the design documents. It is recommended that these reviews occur first at the end of Design Concepts (FEED), the second shall occur during Design Development (Detail) (50 percent), and the third toward the end of Construction Documents Phase (95 percent).
- The CxA compares the design with the interests and needs of GSA as identified in the owner's project requirements. The CxA also compares the proposed design against GSA

Table 7.9 Commissioning agent focused design review scope.

Certification Facilitation	Review contract documents to facilitate project certification goals (i.e. does design meet Energy Star requirements; does Cx meet LEED criteria, etc.)
Commissioning Facilitation	Review contract documents to facilitate effective commissioning (sufficient accessibility, test ports, monitoring points, etc.)
Commissioning Specifications	Verify that bid documents adequately specify building commissioning, including testing requirements by equipment type.
Control System & Control Strategies	Review HVAC, lighting, fire control, emergency power, security control system, strategies and sequences of operation for adequacy and efficiency.
Electrical	Review the electrical concepts/systems for enhancements
Energy Efficiency	Review for adequacy of the effectiveness of building layout and efficiency of system types and components for building shell, HVAC systems and lighting systems.
Envelope	Review envelope design and assemblies for thermal and water integrity, moisture vapor control and assembly life, including impacts of interior surface finishes and impacts and interactions with HVAC systems (blast, hurricane, water penetration).
Fire Protection & Life Safety*	Review contract documents to facilitate effective Cx of fire protection and life safety systems and to aid Fire Protection Engineer in system testing to obtain the GSA Occupancy Permit
GSA Design Guidelines & Standards	Verify that the design complies with GSA design guidelines and standards (i.e. GSA P-100, Court Design Guide, Border Station Guide and Federal Facility Council requirements)
Functionality	Ensure the design maximizes the functional needs of the occupants
Indoor Environmental Quality (IEQ)	Review to ensure that systems relating to thermal, visual acoustical, air quality comfort, air distribution maximize comfort and are in accordance with Owner's Project Requirements.
Life Cycle Costs	Review a life cycle assessment of the primary competing mechanical systems relative to energy efficiency, O&M, IEQ, functionality, sustainability.

(Continued)

Table 7.9 (Cont.)

Mechanical	Review for owner requirements that provide flexible and efficient operation as required in the P-100, including off peak chiller heating/cooling AHU operations, and size and zoning of AHUs and thermostated areas
Operations and Maintenance (O&M)	Review for effects of specified systems and layout toward facilitating O&M (equipment accessibility, system control, etc.).
O&M Documentation	Verify adequate building O&M documentation requirements.
Owner's Project Requirements	Verify that contract documents are in keeping with and will meet the Owner's Project Requirements
Structural	Review the structural concepts/design for enhancements (i.e. blast and Progressive collapse)
Sustainability	Review to ensure that the building materials, landscaping, water and waste management create less of an impact on the environment, contribute to creating a healthful and productive workspace, and are in accordance with Owner's Project Requirements. See also P-100 LEED requirements
Training	Verify adequate operator training requirements

design standards as defined in the latest version of the *PBS* or the company specification in case of applying it in industrial projects like petrochemical or oil and gas projects.

- The CxA identifies any improvements that can be made in areas such as energy efficiency, indoor environmental quality, operations and maintenance, etc. Though the CxA is responsible for reviewing the design from a commissioning perspective, the CxA is not responsible for design concepts and criteria or compliance with local, state, and Federal codes (unless it is specifically called out in their contract).

7.7.2.1.10 Issues Log

All comments and issues identified must be tracked in a formal Issues Log. The Issues Log must be sufficiently detailed so as to provide clarity and a point of future reference for the comments. The Issues Log shall contain the following at a minimum:

- Description of issue
- Cause

- Recommendation
- Cost and schedule implications (on design, construction, and facility operations)
- Priority
- Actions taken
- Final resolution

The Issues Log serves as a vehicle to track, critically review, and resolve all commissioning related issues. The Log is maintained by the CxA and becomes part of the final Commissioning Record.

7.7.2.1.11   Design Review Meetings

- The CxA Team shall have a minimum of three design review meetings (kick-off, concept/DD, and CD). Additional meetings may be required to resolve outstanding issues.
- The CxA is responsible to lead design review meetings and work collaboratively with the Commissioning Team toward the presentation, discussion, and resolution of design review comments.
- Upon resolution of the CxA's comments, the A/E is responsible to incorporate all approved changes into the design documents.

7.7.2.1.12   Update/Refine Commissioning Plan

Now that the Commissioning Agent is on board and has performed Design Stage reviews, the team realigns and updates the Commissioning Plan in preparation for the construction stage.

The Commissioning Team shall formally accept the updated Commissioning Plan before moving into construction. Furthermore, all outstanding comments and issues relative to the CxA's review of the design shall be resolved and accepted changes shall be incorporated into the contract and construction bid documents.

- Commissioning team directory
- Commissioning process activities
- Roles and responsibilities
- Communication structures
- Commissioned systems and equipment
- Commissioning process schedule
- Appendices (owner's project requirements, BOD, Design Review, Issues Log)

7.7.2.1.13   Develop Commissioning Specifications
The commissioning tasks for the contractors will be identified in the commissioning specifications and will include the following:

- General commissioning requirements common to all systems and assemblies
- Detailed description of the responsibilities of all parties
- Details of the commissioning process (i.e., schedule and sequence of activities)
- Reporting and documentation requirements and formats
- Alerts to coordination issues
- Deficiency resolution
- Commissioning meetings
- Submittals
- O&M manuals
- Construction checklists
- Functional testing process and specific functional test requirements including testing
- Conditions and acceptance criteria
- As-built drawings
- Training

Specifications must clearly indicate who is witnessing and documenting the startup of each commissioned system. Specifications must also clearly indicate who is writing, directing, conducting, and documenting functional tests.

The CxA and the A/E must work together to ensure that commissioning requirements are fully integrated and coordinated in the project specifications.

7.7.2.1.14   Written Test Procedures
Written functional test procedures define the means and methods to carry out system/intersystem tests during the construction phase. To the extent possible, these test procedures shall be defined by the Commissioning Team in the Design Stage and written into the contractors' scopes of work.

Test procedures will necessarily be refined early in the construction phase based on the submittal process.

Test procedures provide the following:

- Required parties for the test, which may include the CM, Construction Contractor, specific subcontractor(s), designer, GSA PM, GSA Operating Personnel, GSA Technical

Experts, and Customer Agency representatives. The roles of each required party must also be clearly defined.

- Prerequisites for performing the test including completion of specific systems and assemblies. Prerequisites are of critical importance when undertaking phased construction and/ or phased occupancy. The CxA must coordinate tests with the CM in terms of the overall construction schedule and the anticipated completion of given systems.
- List of instrumentation, tools, and supplies required for the test
- Step-by-step instructions to exercise the specific systems and assemblies during the test. This includes instructions for configuring the system to begin the test and the procedure to return the system to normal operation at the conclusion of the test.
- Description of the observations and measurements that must be recorded and the range of acceptable results

7.7.3   Construction Stage

During the Construction Phase the Commissioning Team works to verify that systems and assemblies operate in a manner that will achieve the Owner's Project Requirements.

The two overarching goals of the Construction Phase are to assure the level of quality desired and to assure the requirements of the contracts are met.

The Construction Phase commissioning activities are a well-orchestrated quality process that includes installation, start-up, functional performance testing, and training to ensure documented system performance in accordance with the owner's project requirements. This testing and documentation will also serve as an important benchmark and baseline for future recommissioning of the facility.

7.7.3.1   Review Submittals for Performance Parameters

As submittals for products and materials are received from contractors, copies of submittals critical to the commissioning process shall be forwarded to the CxA.

In general, the CxA is responsible to review the following types of submittals:

- Coordination drawings
- Redline as-builts
- Product data and key operations data submittals

- Systems manuals
- Training program

Clearly, the CxA cannot review every project submittal. The CxA's review of submittals shall be limited to those items that are critical to the focus of the commissioning process. This review allows the CxA to check the submittals for adherence to the owner's project requirements, basis of design, and project specifications. The CxA shall pay special attention to substitutions and proposed deviations from contract documents and the BOD. The CxA will only comment on submittals to the extent if there is a perceived deviation from the owner's specifications.

7.7.3.2 *Develop and Utilize Construction Checklists*

Construction checklists are developed by the Commissioning Agent, maintained by the Construction Manager, and used by the Construction Contractor and subcontractors.

The intent of construction checklists is to convey pertinent information to the installers regarding the customer agency's concerns on installation and long-term operation of the facility and systems.

The approach to the checklist's structure is to keep it short and simple by focusing on key elements.

Checklists span the duration from when equipment is delivered to the job site until the point that the system/component is started up and operational. This includes testing, adjusting, balancing, and control system tuning.

Checklist Categories

- Delivery and storage checks
- Document and track delivery of equipment and materials to site
- Verify submittal information (avoid accepting and installing equipment that does not meet specifications)
- Ensure equipment or materials remain free of contamination, moisture, and others
- Installation and start-up
- Component-based checks
- Systems-based checks

The CxA will develop the checklist that identifies components and systems for which checklists are required. He or she is responsible for reviewing the owner's project requirements for key success criteria, specifications, and submittals for key requirements. The CxA develops sample checklists for the GSA PM and CM to review, incorporates feedback, and finalizes checklists for distribution.

7.7.3.3   Oversee and Document Functional Performance Testing

Functional performance testing takes over where the construction checklists ended.

The intent of functionally testing the system/building as a whole is to evaluate the ability of the components in a system to work together to achieve the owner's project requirements.

For functional testing to provide valid results, first the individual components and systems have to be verified to be operating properly (see Develop and Utilize Construction Checklists). This includes start-up and testing, adjusting, and balancing (TAB).

The GSA PM must coordinate start-up and installation activities with the GSA.

The Fire Protection Engineer's role in occupancy permitting to include testing for compliance with life safety and code requirements

7.7.3.4   Test Data Records

Test data records capture outcomes of functional performance testing including test data, observations, and measurements. Data may be recorded using photographs, forms, or other means appropriate for the specific test. Test data records shall include, but are not limited to, the following information:

- Test reference (number, specific identifier, etc.)
- Date and time of test
- First test or retest following correction of an issue
- Identification of the systems, equipment, and/or assemblies under the test including location and construction document designation
- Conditions under which the test was conducted (i.e., ambient conditions, capacity/ occupancy, etc.). Tests shall be performed under steady-state and stable conditions.
- Expected performance
- Observed performance including indication of whether or not this performance is acceptable
- Issues generated as a result of the test
- Dated signatures of those performing and witnessing the test

7.7.3.4.1   Test Issues and Follow-up

The functional performance tests are the heart of the commissioning process and they are also the most difficult and time consuming. System troubleshooting is a critical function of the CxA.

As inspecting and testing proceed, despite the team's best efforts, the CxA will find a number of items that do not appear to work as intended. There will be a certain amount of system retesting that will be performed by the CxA because of system deficiencies during the initial testing.

In order to assure success, the GSA PM shall allow some time in the schedule and money in the budget for retesting. The GSA PM shall be apprised that issues resolution and associated financial implications are a common point of contention between parties.

If equipment or systems are found to be malfunctioning, these problems shall be documented and listed in the Issues Log for team resolution. The Issues Log must be very clear about the test, system(s) involved, and tracking of the problem as it is corrected.

Both the amount of retesting paid for by GSA versus the amount paid by the contractor and/or designer, as well as the parameters for which parties are responsible for correcting deficiencies, shall be very clearly spelled out in the contracts.

7.7.3.5   Hold Commissioning Team Meetings and Report Progress

Consistent, regular Commissioning Team meetings are essential to maintain the progress of the project and the momentum of the commissioning process.

The schedule of meetings shall be defined, documented, and included in appropriate bid documents during the Design Stage (monthly construction phase CxA Team meetings are recommended).

Team members at meetings must be authorized to make commitments and decisions for their respective parties. The typical agenda for construction phase Commissioning Team meetings shall include items such as previous action items, outstanding issues, schedule review, new issues, etc.

In addition to regular meetings, the CxA is responsible for preparing monthly commissioning process reports during the construction phase. These reports shall include, at a minimum, the following information:

- Progress and status report along with look-ahead
- Identification of systems or assemblies that do not perform in accordance with the owner's project requirements

- Results from latest version of the Issues Log (importance, cost, and measures for correction)
- Test procedures and data
- Deferred and seasonal tests (and reason for deferring)
- Suggestions for enhancements that will improve the commissioning process and/or the delivered facility

7.7.3.6 Conduct Owner Training

An important step in the commissioning process is ensuring that GSA Operating Personnel are properly trained in the required care, adjustment, maintenance, and operations of the new facility equipment and systems. It is critical that operations and maintenance personnel have the knowledge and skills required to operate the facility to meet the owner's project requirements. Training shall specifically address the following:

- Step-by-step procedures required for normal day-to-day operation of the facility
- Adjustment instructions including information for maintaining operational parameters
- Troubleshooting procedures including instructions for diagnosing operating problems
- Maintenance and inspection procedures
- Repair procedures including disassembly, component removal, replacement, and reassembly
- Upkeep of maintenance documentation and logs
- Emergency instructions for operating the facility during various nonstandard conditions and/or emergencies
- Key warranty requirements

Because of the Commissioning Agent's in-depth knowledge of the design intent and building systems, it is important to have the CxA intimately involved in the training.

The CxA is responsible for facilitating the entire owner training process. This process begins in the Design Stage by assuring that appropriate levels of training are planned and included in the specifications.

The CxA maintains a system-based, as opposed to component-based, focus in the training to ensure that operating personnel understand the interrelationships of equipment, systems, and assemblies. The CxA also reviews agendas and material developed by the contractors in advance of the trainings for quality, completeness, and accuracy.

The CxA shall also attend a number of the key training sessions to evaluate effectiveness and suggest improvements in the delivery of the material.

The majority of training shall be done during the construction phase prior to substantial completion. Some systems and assemblies may require ongoing training during occupancy and post-construction. The exact systems, subsystems, equipment and assemblies that require training, as well as the required number of hours of training, are spelled out in the project specifications. The CM utilizes attendee sign-in sheets to verify that the training was delivered to the intended staff.

The instruction shall be given during regular work hours (for all shifts) on such dates and times that are selected by the GSA Project Manager. The instruction may be divided into two or more periods at the discretion of the GSA PM.

It is highly recommended that all trainings be videotaped. Videotaping trainings allows for future reference of the material and training of new employees down the road.

The team may also wish to consider DVDs in lieu of videotapes for reasons of longevity and convenience.

The Contractor shall be required to provide the GSA PM with an edited draft version of the taped training sessions (generally within seven days), which include all aspects of the operation, inspection, testing, and maintenance of the systems.

Technical experts will review the videotape and provide the contractor with comments. The contractor will then resubmit an edited final version of the tape (generally within seven days of receipt of comments).

Instructor Qualifications

- The instructor shall have received specific training from the manufacturer regarding the inspection, testing, and maintenance of the system provided. The instructor shall train the government employees designated by the contracting officer in the care, adjustment, maintenance, and operation of the new facility equipment and systems.
- Each instructor shall be thoroughly familiar with all parts of the installation. The instructor shall be trained in operating theory as well as practical operation and maintenance work.

7.7.3.7 Turnover Commissioning Record

It is critical to understand that commissioning documentation is developed throughout the project and turned over before substantial completion.

Table 7.10. Commissioning record document.

Document	Phase Started	Developed/Provided By
Commissioning Plan	Planning	GSA PM
Commissioning Plan Appendices		
A. Owner's Project Requirements	Planning	GSA PM
B. Basis of Design	Design	A/E
C. Commissioning Specifications	Design	A/E/CxA
D. Design Review	Design	CxA
E. Submittal Review	Design	CxA
F. Test Procedures	Design	CxA
G. Issues Log	Construction	CxA
H. Construction Checklists	Construction	CxA/Construction Contractor
I. CxA Site Visit & Cx Team Mtg. Minutes	Construction	CxA
J. O&M Review	Construction	CxA
K. Training Documentation	Construction	CxA/Construction Contractor
L. Warranty Review	Construction	CxA
M. Test Data Reports	Construction	CxA/Construction Contractor
Summary Report	Construction	CxA
Recommissioning Management Manual	Construction	CxA/GSA PM

Commissioning documentation turned over at this stage of the project is a result of a well thought out documentation plan and collection of information throughout all of the project phases.

Table 7.10 outlines necessary documentation of the commissioning process by project phase in order to complete the commissioning.

The Commissioning Record shall include a brief summary report that includes a list of participants and roles, a brief building description, overview of commissioning and testing scope, and a general description of testing and verification methods. For each piece of commissioned equipment, the report shall contain the disposition of the CxA regarding the adequacy of the equipment, documentation, and training meeting the contract documents in the following areas:

1. Equipment meeting the equipment specifications
2. Equipment installation
3. Functional performance and efficiency
4. Equipment documentation
5. Operator training

The Recommissioning Management Manual provides guidance and establishes timelines for recommissioning of building facilities systems and components. The format of the Recommissioning Management Manual will closely parallel the Commissioning Plan for the facility.

7.7.4 Building Commissioning Process Post-Construction Stage

7.7.4.1 Post-Construction Stage

Systems, assemblies, equipment, and components will tend to shift from their as-installed conditions over time.

In addition, the needs and demands of facility users typically change as a facility is used.

The post-construction stage allows for the continued adjustment, optimization, and modification of building systems to meet specified requirements.

The objective of the post-construction stage is to maintain building performance throughout the useful life of the facility.

The active involvement of the Commissioning Agent and the Commissioning Team during initial facility operations is an integral aspect of the commissioning process.

Commissioning activities during Post-Construction include issues resolution, seasonal testing, delivery of the Final Commissioning Report, performing a post-occupancy review with the Customer Agency, and developing a plan for recommissioning the facility throughout its life cycle.

7.7.4.2 Perform Deferred and Seasonal Testing

Due to weather conditions, not all systems can be tested at or near full load during the Construction Phase.

For instance, testing of a boiler system might be difficult in the summer and testing of a chiller and cooling tower might be difficult in the winter. For these reasons, commissioning plans shall include offseason testing to allow for testing of certain equipment under the best possible conditions.

In addition to seasonal testing, several systems may have been deferred during the initial testing for a number of reasons including prerequisite activities not complete, phased occupancy issues, and improper testing conditions.

The commissioning team must use the Issues Log as a guide during the post-construction stage to complete all deferred testing.

7.7.4.3   Re-inspect/Review Performance Before End of Warranty Period

During the first year of the building's operation, it is important to assure that the performance of the facility is maintained, particularly before the warranty period expires.

Ten months into a twelve month warranty period, the operation of system and components is critically reviewed by the CxA, owner, and CM to identify any items that must be repaired or replaced under warranty.

This review is based on warranty items and continued performance with the owner's project requirements.

Discrepancies between predicted performance and actual performance and/or an analysis of any complaints received may indicate a need for minor system modifications.

The CxA documents the results and forwards recommendations to Owner and CM for resolution.

The GSA PM shall be cognizant of the impacts of a phased occupancy, if applicable, on the warranty period and make necessary adjustments for review and inspection.

Proper maintenance programs, training, and familiarization of the systems by the new operating staff are important to support post-construction commissioning.

For example, a standard method of recording and responding to complaints must be in place and used consistently.

As equipment and controls are replaced throughout the maintenance program, calibration and performance must be checked, documents revised, and any changes or new equipment data sheets included in the operations and maintenance manuals.

7.7.4.4   Complete Final Commissioning Report

During Post-Construction, the CxA is responsible for delivering a Final Commissioning Report. This document is additive to those items detailed in the Turnover Commissioning Record. The Final Commissioning Report shall include the following, at a minimum:

- A statement that systems have been completed in accordance with the contract documents and that the systems are performing in accordance with the final owner's project requirements document
- Identification and discussion of any substitutions, compromises, or variances between the final design intent, contract documents, and as-built conditions
- Description of components and systems that exceed owner's project requirements, those which do not meet the requirements and why, summary of all issues resolved and unresolved, and any recommendations for resolution
- Post-construction activities and results including deferred and seasonal testing results, test data reports, and additional training documentation
- Lessons learned for future commissioning project efforts
- Recommendations for changes to GSA standard test protocols and/or facility design standards

The Final Commissioning Report will serve as a critical reference and benchmark document for future recommissioning of the facility. In addition, the CxA is responsible at this stage to assure the engineering office is completely updated with the as-built drawings.

7.7.4.4.1   Final Satisfaction Review with Customer Agency
The GSA PM will lead a final satisfaction review with the Customer Agency. This review shall occur at one year after occupancy.

Minimum attendees shall include the Commissioning Team and other selected Customer Agency representatives.

The purpose of this review is to obtain honest, objective, and constructive feedback on what worked well throughout the commissioning process and what the Commissioning Team could have done better. The group shall be focused on identifying root causes and proposing corrective action for future projects. Specific discussion topics may include the following:

- Owner's Project Requirements
- Systems selected for commissioning
- Coordination issues
- Commissioning budget and costs
- Commissioning schedule relative to project schedule
- Occupant comments/complaints
- Documentation issues
- Lessons learned

The GSA PM takes the lead on documenting this session in a formal lessons learned report. This information will be an important input to future projects.

7.7.4.5 Recommission Facility Every 3–5 Years

At this stage of operation, a considerable investment has been put into assuring the facility operates as intended. Understanding that systems tend to shift from their as-installed conditions over time due to normal wear, user requests, and facility modifications, it is strongly recommended that Customer Agencies consider recommissioning facilities every three to five years.

A facility recommissioning program serves to assure operational efficiency and continued user satisfaction. Maintaining good O&M and occupant complaint records is key to continued recommissioning efforts.

7.7.4.6 Recommissioning

Recommissioning shall generally include the following:

- Establishing that the original basis of design and operation is still appropriate for use, occupancy, tenant agencies, and GSA goals and modify the operations/controls sequencing as appropriate for optimum operations
- Reviewing and benchmarking key systems operations/performance against the Basis of Design
- Evaluating envelope tightness/pressurization by infrared or other methods
- Performing energy analysis
- Recommending repairs/modifications to optimize building performance

It is important to recognize that at three to five years after occupancy, the GSA PM will likely not still be involved with a particular project. Therefore, the Customer Agency will take the lead on facility recommissioning.

Recommissioning shall include Commissioning Agent services. While there are obvious benefits of familiarity, the Customer Agency may or may not bring back the project Commissioning Agent. Recommissioning is not part of the original CxA's contract and, therefore, the Customer Agency must procure these services through a RFQ/RFP process at the time of recommissioning.

7.7.5   Advantages for Total Building Commissioning System

This system was applied to different administration buildings. Table 7.11 presents the cost for commissioning a retrofit project. Table 7.12 shows the cost for applying building commissioning to existing buildings. One can see that the cost payback varies from 0.3 to 1.6 years.

Although little research has been completed to document the link between comfort and productivity in the office environment, comfortable employees are generally considered to be more productive than uncomfortable employees.

When occupants of an office building complain of discomfort, additional costs and lost productivity have been estimated to be significant.

One such estimate assumes that the typical building has one occupant per 200 square feet of space and an annual payroll cost of $34,680 per person (or $173 per square foot of office space).

If one out of every five employees spends 30 minutes a month complaining about the lighting, the temperature, or both, the employer loses $0.11 per square foot in annual productivity. In a 100,000-square-foot building, this loss amounts to $11,000 per year.

The comparison of cost savings in energy by applying the building commissioning system for an existing building in 1995 is presented in

Table 7.11 Costs for commissioning retrofit project.

Scope	Cost range
Total building commissioning	0.5–1.5% of total construction contract cost
HVAC and automated control system	1.5–4% of mechanical contract cost
Electrical system	1–1.5% of electrical contract cost

Table 7.12 Existing building system commissioning cost.

Building type	Commissioning cost (USD in 1995)	Annual savings (USD in 1995)	Payback cost (years)
Computer facilities/office	24000	89758	0.3
High rise office	12745	8145	1.6
Medical institute	24768	65534	0.4
Retail	12800	8042	1.6

Table 7.13 Annual energy and cost savings from commissioning existing building systems projects.

Building type	Energy savings	Cost savings (USD in 1995)
22000 ft2 office	130800 KWh	7736
110000 ft2 office	279000 KWh	12447
60000ft2 high-tech manufacturing facility	336000 KWh	12168

Table 7.13, noting that the cost saving is about \$12,168 for 60,000ft^2 of manufacturing facilities.

Quiz

1. Why is quality planned and not inspected?
   - It reduces quality and is less expensive.
   - It improves quality and is more expensive.
   - It reduces quality and is more expensive.
   - It improves quality and is less expensive.

2. What is meant by your product or service completely meets a customer's requirements?
   - Quality is achieved.
   - The cost of quality is high.
   - The cost of quality is low.
   - The customer pays the minimum price.

3. All of the following are part of quality control EXCEPT?
   - Cost of quality
   - Inspection
   - Control charts
   - Flowcharting

4. A project is in progress and the project manager is working with the quality assurance department to improve stakeholders' confidence that the project will satisfy the quality standards. Which is the gain from this process?
   - Quality problems
   - Checklists
   - Quality improvement
   - Quality audits

5. While testing the strength of concrete poured on your project, you discover that over 35% of the concrete does not meet your company's quality standards. You feel certain the concrete will function as it is and you don't think the concrete needs to meet the quality level specified. What should you do?
   - Change the quality standards to meet the level achieved.
   - List in your reports that the concrete simply "meets our quality needs."
   - Ensure the remaining concrete meets the standard.
   - Report the lesser quality level and try to find a solution.

6. A project manager says to a team member, "If you cannot complete this task according to the quality standards you set in place, I will remove you from the team that is going to Paris for the milestone party with the customer." What form of power is the project manager is using?
   - Reward power
   - Formal power
   - Penalty power
   - Referent power

8

Practical Risk Management for Oil and Gas Projects

8.1 Introduction

In Chapter 2, risk assessment for a project is discussed from an economic project of view. In this chapter we will focus on defining, controlling, and mitigating the risk during project execution, especially in the case of construction on site. However, this risk assessment technique can also be used at the last stage of the engineering.

When studying risk assessment for a project from an economic point of view, the probabilistic studies and Monte-Carlo simulation techniques are key in this assessment, which is called quantitative risk assessment. But, this method of analysis and assessment requires special software with specialized skills. We will discuss applying risk assessment in the execution phase using qualitative risk assessment tools in detail in this chapter. This assessment method does not require a special skill set or software, as the main player in this method is the project team members with a project leader or facilitator who have experience with the qualitative risk assessment in similar projects.

The risk management of a project is the process of identifying risks that could affect the project and how to control them. We must trust that we can overcome these risks well, if they are studied scientifically.

In general, risks are events that, if they occur, would affect the outcome of a project and, consequently, not satisfy the stakeholder. To be specific, there are events or activities that can be performed in a poor manner, thus increasing the project cost or time or affecting the quality.

As shown in Figure 8.1, the uncertainty is like the black box, in which no one can know what will happen. So, we can say that objectives are things that must happen, while uncertainties are things that might happen.

Risk management is an ongoing process during the implementation of a project until the end of that project. The potential risk is substantially less at the completion of the project.

The risks can be classified in two categories:

- Project risks are the risks that can happen during a project due to technical mistakes that can occur during construction.
- Process risks are the risks that can occur during the project due to procedural mistakes, poor communication between the project team, or poor team performance.

In general, there are many sources of uncertainties, especially in the main elements of a project, which are cost, time, quality, and HSE as presented in Figure 8.1. Our target is to control these uncertainties, try to predict what could happen, and avoid it in a reasonable time.

8.1.1   The Risk Management Process

The Project Management Institute (PMI) uses the systems approach to risk management found in the *Guide to the PMBOK*. The risk process is divided into six major processes:

1. Risk management planning
2. Risk identification
3. Risk assessment
4. Risk quantification

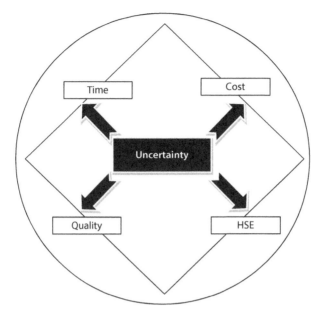

Figure 8.1 Sources of Uncertainty.

5. Risk response planning
6. Risk monitoring and control

These major processes will be discussed in the following section.

8.2   Project Risks

After completing the time schedule, the potential risk will be more obvious. Knowledge of the risks that may be faced during the project is extremely important for the project manager, as he or she is responsible for identifying the activities of higher risk impact on the overall project implementation, which will either increase the duration or the cost.

Therefore, the project manager should review the planning schedule and identify areas of planning that contain high risks, as known from the following:

1. Tasks on the critical path
2. Tasks that need a long time period in which to be executed
3. Tasks that have little overtime
4. Activities that start with the beginning of other activities

5. Tasks that need many individuals for their execution
6. Complex tasks
7. Activities and tasks that need condensed training
8. Tasks that need new, advanced technology

After you select the tasks that would cause risk to the project, then you will need to identify and plan the necessary steps to implement those tasks and how to follow up on implementation daily and assign reasonable persons who will be responsible for follow-up in that stage of the project.

In order to have a sense of high-risk activities, let us consider the example of pouring concrete in Chapter 3 and answer the following question: what are the risk activities?

One can find that the highest risk activity is the excavation, as it needs a long duration in execution and it is on the critical path. Therefore, it has a high probability of delay and, consequently, it has a high potential impact on the complete project schedule.

In any project, the longest time period activity located on the critical path is the most risky activity in the project. On the other hand, the machines and mechanical equipment are frequently coming from outside the project country, so the delivery of the equipment is most critical and an expected risk is high. In addition to that, in most cases, many other activities depend on the delivery of this equipment onsite.

These activities have a high-risk assessment, as they will be out of full control. The probability of delay is high and also has a direct impact on the project completion time.

The success of the project means that the project team succeeded in achieving the objectives of the project on a specific time schedule and budget. It is known that there is nothing specific in nature. For example, specific costs, the time period, and the objectives of the project can increase and decrease. It is worth mentioning, that these three elements affect each other, so the success of the project requires each element of the project to work cohesively, as presented in Figure 8.2.

Per the probability theory that was discussed in Chapter 2, the probability of success is small because all the occurrences of achieving the project objectives within the time and cost should happen. Therefore, our goal is to locate our project in this zone of intersection.

There are many areas in the project that are not specific and these are sources of risks, which can be any of the following:

- Activities of a long period of time and on the critical path
- A lack of identification of the project objectives

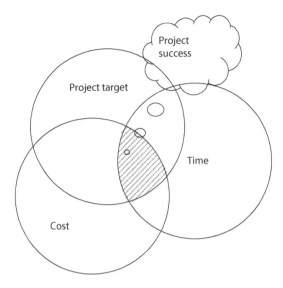

Figure 8.2 Point of project success.

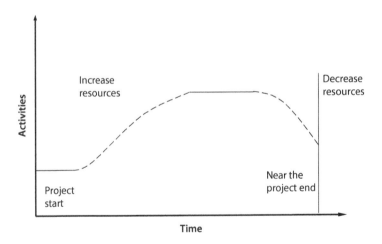

Figure 8.3. Change in staff volume during a project.

- A non-competent project manager
- An inaccurate cost estimate
- A bad atmosphere, in general, in the project
- Achieving customer satisfaction
- A rapid change in resources during time periods, as shown in Figure 8.3

Figure 8.3 presents staff and other resource distribution onsite within a construction period. It is shown that, at the start of the project, if the activities are low as well as the resources, then it will be a transition zone to increase the resources consequently with increasing the activities. This transition zone will be a high risk, as it will rapidly increase the working resources in the project in a short time period. Therefore, the likelihood of bad quality, misunderstanding the objective, and mistakes in safety procedure will be very high.

In the middle of the project, there will be stability in the number of resources, so the risk will be less. After that, start the other transition zone by demobilizing the resources, which will be a high risk also as, in this case, there will be a likelihood of manpower shortage and missing the handover or transfer of equipment by mistake. In the same time, the staff decreasing so everyone in the project will be busy searching for jobs in other projects.

The list below can be your guide as a checklist for defining uncertainty risks in your project. In general, the common sources of uncertainty in a project are as follows:

- Scope of work
- Quality of estimates
- False assumptions
- Technological novelty
- Changes in technical specs
- User interface
- Staffing
- Staff productivity
- Skill levels
- Contractor performance
- Subcontractor performance
- Approvals and funding
- Market share
- Competition
- Economic climate
- Inflation and exchange rate
- Site conditions, such as soil characteristics
- Weather, as it has a high impact in the case of offshore projects
- Transportation logistics
- Change in law
- Political environment

- Public relations
- Customers
- Extensive software development

8.3   Risk Assessment

The risk assessment procedure is shown in Figure 8.4. The first step is to define the expected risks during the project execution and then analyse this risk. The last step is to prioritize these risks.

Each project has a risk, no matter what the project is. We focus on the risks affecting the management of the project and, by knowing these risks, we can set priorities to develop solutions for those risks and to mitigate them.

In order to assess these risks, we must answer the following questions accurately and impartially:

- What is the risk exactly?
- How do these risks affect the project?
- What can be done to reduce the impact of the risks?

Now we will discuss a method for determining the risks in a different way than mentioned previously. We previously explained assessing the risk of a project as a whole by using a quantitative risk assessment methodology as a part of the feasibility study.

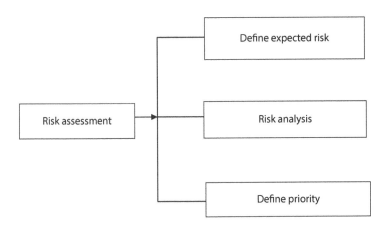

Figure 8.4. Risk assessment tools.

In this stage, the risks will be assessed by their effects on the objectives, time, and cost. Now we need an easy way to assess the risks practically. This method is called "qualitative risk assessment."

In the case of oil and gas projects specific for offshore structures, the risk is very high to meet project time and cost as the construction offshore depends on the marine vessels, whose performance is affected by the weather. In the case of probability of high weather, it means the vessel must stop working, which means delay in time in addition to the accumulation of a standby fee for the vessel.

8.4   Risk Identification

The identification of risks is very important. Each item must be described in detail so that it will not be confused with any other risk or project task that must be done. Each risk should be given an identification number. During the course of the project, as more information is gathered about the risk, all of this information can be consolidated.

Initially, you want to determine the risks and you can do that professionally by inviting the working team members in the project to a special meeting to determine the potential risks of the project.

It is convenient to hold the meeting at an offsite location, such as in a hotel or another meeting room out of the work environment. This is mainly for two main reasons: first, the team has to start working together so you need discussion away from work stress to melt the ice; second, you need their minds free of any outside stress so they can provide knowledge of their previous experiences in predicting the probability of events occurring, which depends mainly on the back history.

At this meeting, use the method of brainstorming so that each individual gives his or her own thoughts regarding expected risks during the project. Then, collect all these ideas and write them in the order as in the following table, Table 8.1. This table will be the basic document of the project.

The first component we need to discuss is the identification of the risk event. In the course of identifying risk events, we will call upon the project team, subject matter experts, stakeholders, and other project managers. Much of the work already done in the project will be utilized in the risk management process. Among with these items that will be used are the project charter, the work breakdown structure, project description, project schedule, cost estimates, budgets, resource availability, resource schedules, procurement information, and assumptions that have been made and recorded.

Table 8.1. List of Project Risk.

Project name							
Project manager:							
Client name:							
Responsible	Risk degree: High Medium lowv	Date	Impact: High/medium/low	The probability: 1-9	Risk description	Key stage code	Item No.
Project manager signature:						date:	

8.4.1 Methods of Defining Risk

There are many ways to discover and identify risks. We will discuss several of them here, but, to be clear, brainstorming is the most traditional and practical method. However, the others can also be used in special circumstances:

- Brainstorming
- Delphi technique
- Nominal group technique
- Crawford slip
- Expert interviews
- Root cause identification
- Strengths, weaknesses, opportunities, and threats (SWOT) analysis
- Checklists
- Analogy
- Documentation reviews

8.4.1.1 Brainstorming

Brainstorming is probably the most popular technique for identifying risks. It is useful in generating any kind of list by mining the ideas of the participants. To use the technique, a meeting is called to make a comprehensive list of risks. It is important that the purpose of the meeting be explained clearly to the participants and it is helpful if they are prepared when they arrive at the meeting. The meeting should have between ten and fifteen participants. Try to avoid having less than ten participants because there might be less interaction among the participants.

If there are more than fifteen people, the meeting tends to be difficult to control and keep focused. The meeting should take less than two hours.

For larger projects, it may be necessary to hold several meetings. Each meeting should deal with a separate part of the project and with the risks associated with that project part. This will keep the number of persons involved to a reasonable size and the meetings will be much more productive.

It's best if the facilitator is not an internal employee or is from a different department, but he or she must have experience for this type of project.

In the beginning, the facilitator should illustrate the project briefly and the illustration chart should be hanged on the wall and should contain a time schedule with the main key stages. In addition, the facilitator should define the stakeholder, the budget, and the location of the project.

Then, the facilitator and the team leader direct the dialogue to start asking about the potential risks. They suggest that every expected risk defined should be put on a list to be shown to everyone. In this case, there will be no discussion for any item. After putting the entire expected potential list together, begin the next step but take care from the following:

- Do not stop or reject any idea.
- Never let anyone suggest to remove an idea.
- Do not allow interruption in the discussion.
- Do not go deeply into detail.

Then, start to filter these risk items and merge any two or three risks with a general one if similar.

Then, put them in Table 8.1. Define the likelihood of an event occurring during a project and, from the table, also define the consequence. Also, you will have three alternatives to take the consequence that most of the participants agree on. The other column is the manageability of this item. In the case of obtaining approval or permits from the government, it will be low manageability as it is out of your control, but if the item of risk is to change a drawing or to do an As-Built drawing, it will be easy to manage.

8.4.1.2   Delphi Technique

Brainstorming is not very ideal in cases where the attendees are not familiar with risk assessment technique. Also, if there is a high level manager in the meeting room, everyone may try to only present one's self and may not freely provide ideas.

In many cases, the meeting output is not always 100 percent efficient. In some cases during the project, you may face potential risks that were not considered in the meeting. This means all the effort was gone. So this meeting should be held in a professional manner to cover any potential risk or this meeting will be a waste of time and effort.

The Delphi method is not traditional in practical life, but it can be used. The name "Delphi" is derived from the Oracle of Delphi. The authors of the method were not happy with this name because it implies "something oracular, something smacking a little of the occult." The Delphi method is based on the assumption that group judgments are more valid than individual judgments.

The Delphi method was developed at the beginning of the Cold War to forecast the impact of technology on warfare (1999). In 1944, General Henry H. Arnold ordered the creation of the report for the U.S. Army Air Corps on the future technological capabilities that might be used by the military.

Different approaches were tried, but the shortcomings of traditional forecasting methods, such as theoretical approach, quantitative models, or trend extrapolation, in areas where precise scientific laws have not been established yet, quickly became apparent. To combat these shortcomings, the Delphi method was developed by Project RAND during the 1950s and 1960s (1959) by Olaf Helmer, Norman Dalkey, and Nicholas Rescher (1998). It has been used ever since, together with various modifications and reformulations such as the Imen-Delphi procedure.

This method is based on the idea that the opinion of a group is better than the opinion of one person. As in the war, this method was used by asking experts to give their opinion on the probability, frequency, and intensity of possible enemy attacks. Other experts could anonymously give feedback. This process was repeated several times until a consensus emerged.

In the same way, ask every expert in this type of project separately about the expected potential risk of the project. Now days, this method is easy due to e-mail, video conferences, and other methods of communication that appear every day. The process begins with the facilitator using a questionnaire to solicit risk ideas about the project. The responses from the participants are then categorized and clarified by the facilitator. After reviewing the response from the participant and categorizing it, the facilitator rounds these risks to the participant and then does it for another round until they settle the final list of project risks.

The largest benefit from this method is that the cost is very little or there is no cost, but it takes a great effort from the facilitator.

8.4.1.3   Nominal Group Technique

The nominal group technique has the same condition as the brainstorm meeting, but in this technique the facilitator requests from everyone in the meeting room to write the expected risks during the project on a piece of paper. Noting that, the number of participants is usually seven to ten people.

When this is completed, the facilitator takes each piece of paper and lists the ideas on a flip chart or blackboard. Here, we avoid the disadvantages of the brainstorming technique. Until this point in the process no discussion has taken place.

After all the ideas are listed on the flip board, start the discussion, explanation, and clarification for every item. Then, compile the risks that are similar.

This process reduces the effect of a high-ranking person in the group, but does not eliminate it like the Delphi technique does. The nominal group technique is faster and requires less effort on the part of the facilitator than the Delphi technique.

8.4.1.4 Crawford Slip

The Crawford slip process has become popular recently. The procedure of the meeting is the same as the brainstorming technique. The advantage of this method is that it can be used for more than ten participants. The Crawford slip process does not require as strong a facilitator as the other techniques and it produces lots of ideas very quickly. This method depends on the facilitator asking the question as "what is the expected risk in the project?" and every one writing the answer on separate pieces of paper and then repeating this question another ten times. If you have ten people, you will obtain about 100 answers to the same question. Usually, there will be duplicate answers, so filter all the answers and put them in a list to be shown to everyone in this meeting.

The Crawford slip meeting can take place in less than half an hour for a small group, but the time increases with an increase in the number of participants.

8.4.1.5 Expert Interviews

We can use an expert in these types of projects. It's preferable that the expert works for the company because then he or she would have information about the project in addition to being able to easily imagine the project constrains and potential risk.

For instance, if the procurement employee in your company is not competent, he will feel that, but if the expert is from outside the company, you will be afraid to transfer him information about the competence of the procurement in your company or be afraid to tell him that the management has hesitated to make decision or that the paper work takes a lot of time inside the company. Another major benefit of using experts from within your company is that it will be free of charge, whereas external experts will cost.

Before the interview, you should collect all the data about the project and present it clearly to the expert to let him advise on the most probable risks in the project.

In addition, the goals of the interview must be clearly understood. During the interview, the information from the expert must be recorded. If more than one expert is used, the output information from the interviews should be consolidated and circulated to the other experts.

8.4.1.6 Root Cause Identification

This method is used in conjunction with the brainstorming technique and other similar meetings, as the facilitator should be able to define the root cause in any item.

This method is based on root cause analysis (RCA). The main target of the RCA is to answer the questions of what, how, and particularly why something could go wrong in the project.

This method researches deeply into the event probability and consequence in addition to the main cause that let this event happen.

For example, assume that there is a delay in performing a job offshore due to weather. This method needs to define the way to obtain the weather forecast. For another example, assume the potential risk is that spare parts will be delivered late on the offshore platforms. As we investigate why this happened, we find that there was a high wind speed and the transfer of the parts was delayed for five days. The root cause of the problem is not the high wind speed, but rather that ordering the mechanical spare parts should be done early enough to allow for the likelihood of bad weather.

8.4.1.7 Checklists

Checklists have gained popularity in recent years because of the ease of communicating through computers and the ease of sharing information through databases. Every company has established checklists based on their experience and examples of some of the risks are presented in section 8.2. The checklist is a concrete thinking method, but when predicting risk, it needs a freethinking method.

8.4.1.8 Documentation Reviews

This method depends on reviewing lessons learned from previous projects. Take care to close out reports for the projects which contain lessons learned for more than twenty years after the ISO effectively started in the companies and that their staff follow PMP principal in the projects. In addition, your company should have performed this type of project beforehand many times and the company should have a strong database containing learned lessons from projects. In any account, it is very important as it is zero-cost. If you have a small project, you can start the risk assessment meeting that is described above.

Other methods can be used in defining the risk, but they are not as practically used as the SWOT analysis method, which is a diagram method that depends on providing a flowchart that shows the sequence of events that take place within a given time.

The analogy method is part of the document review, as this method depends on obtaining the risk management plan of other projects that were similar and an analogy can be formed. By comparing two or more projects, you can find similar characteristics for each project and, from that, we will have an overview of the risks of the new project.

8.4.2   Grouping the Risks

After defining all the events that may occur during project execution, affect the project objective, or increase the project time or cost, you then need to group the risks. In this stage, we need an applicable method to identify the risks. This method is called quantitative risk assessment. This method is the same as defining the key stage. All participants in the project should attend a meeting, in which the brain storming technique is used to predict events whose occurrence make a problem to the project target, cost, and time.

Grouping the risks will be more important for large projects than small ones. The general idea is that if it takes more than ten people to meet and deal with a group of risks, the meeting is too large and will be inefficient. As projects become larger, it is necessary to have a series of risk management meetings, whereas in a small project one meeting might do. To facilitate this, you can use techniques similar to the techniques that were used in the development of the work breakdown structure. In fact, the WBS itself can be used to organize meetings for risk management.

Risks should be assigned to the person who is most closely associated with where the risk will have its largest impact or to the person who has the most familiarity with the technology of the risk. A risk that takes place during the completion of a particular task and directly affects only that task should be a concern to the person responsible for that task. Of course, no task in a project is truly independent of all the others. So for more severe risks, a person in the organization above the person responsible for the task may be responsible for the risk.

Oftentimes, in projects where risk is of great concern, the project manager creates the position of risk manager. This person is responsible for tracking all risks and maintaining the risk management plan. As projects become larger or tolerance for risk is low, this approach becomes more necessary.

The responsible person who will mitigate and follow up this task should be defined clearly, and his or her name can be placed in Table 8.3.

8.5 Define Priorities

There are many ways to define the risk ranking for each risk item that was defined before. In this step, the facilitator will ask the team in the meeting to define the degree of likelihood that the event will occur and, also, define how much it can impact the project. There are two traditional methods. The matrix method is the most popular and is defined in the PMOBOK guide. The other way depends on ranking the risks in a table and defining them in three categories: red, yellow, and green for every identified risk.

8.5.1 Matrix Method

In the previous meeting everyone wrote down all expected risks. Now you want to determine the priorities of those risks. We will use this experience in assessing the risk at the same meeting and there will be a two-stage evaluation process.

The first phase requires determining the likelihood of a risk and rating it from one to nine, one being the least likelihood of the event occurring and nine being the greatest probability of the event occurring.

In the second stage, the team determines the outcome of an event on the project or the expected losses due to this event occurring and it is divided into the following cases:

- High: It has a greater significant impact on the time schedule or the cost of the project.
- Medium: It has a medium impact on the time schedule and cost.
- Low: It has a lower influence on the time schedule and little impact on the cost.

In summary, the following table can be used to determine the risk priorities.

The team must know what the probability of the event occurring is and what its impact is to the project. The impact will be obtained through the discussion for each event by answering the two questions: What is the probability of potential risks and what is their impact on the project?

From the previous matrix shown in Table 8.2, one can determine the risk category.

In the case of an unacceptable risk, the event needs to be analysed with high accuracy and a thorough examination or by focusing on finding solutions to them. If their occurrence will cause failure for the whole project and the project cannot be done as planned, the risks are unacceptable and must be resolved now.

In the case of high-risk, this situation occurs when the event could have a significant impact on the project schedule and cost. Therefore, sustained monitoring is required.

In the case that a medium-risk event occurs, medium impact is expected and not all the key points of the project will be affected. Therefore, the risk mitigation must be reviewed at each meeting of the project and evaluate their action work and follow it up periodically.

In case of a low-risk event, it is not expected at the time of the event to have a significant impact on the project and must follow up from time to time.

The risks that have been assessed as high risk at the follow-up may cause some changes due to some work or actions performed, which can reduce or increase of the probability of the event occurring.

After identifying the risks, rank them from highest to lowest risk.

In every item, find a solution and accurately identify the responsibilities as to who will do the activity. This will reduce the risk of the event and it will be registered, as shown in Table 8.3.

8.5.2 Tabulated Method

The recognized method for the calculation of risk is the probability of the event occurring with knowledge of the impact of that event on the project

Table 8.2 Risk Ranking Score.

Probability of event occurring		Impact on the project		
		Low	Medium	High
	7–9	Medium	High	Not acceptable
	4–6	Low	High	Not acceptable
	1–3	Low	Medium	High

Table 8.3 Risk Management.

Project:	
Project Mgr.:	
Risk Number:	Risk title:
Risk description:	
Risk degree:	Probability: 1 2 3 4 5 6 7 8 9 Impact: High Medium Low
Project areas that will affect by the event:	
Ordering of the event impact on the project:	
Who is responsible:	Solution step:
Prepared By:	
Project Mgr. Sign.: date:	

as a whole. The importance is calculated by means of monetary value and, from that, the degree of the event impact on the project will be defined.

Some modern techniques are inevitable to take into account and the ability to manage the event and to understand that there are some pitfalls that can be solved easily and managerially.

In this method, which is more practical, managing the activity is very important for the risks. So, there is a column in the table to define the manageability.

For example, discrepancies between departments can be solved by holding meetings between departments from time to time. Therefore, in this event it is very easy to manage, so give this risk a lower score (to decrease the whole risk score value) for the manageability of the event.

On the other hand, if the problem is related to contact outside organization, such as a deal with the government for license or permissions or other, you do not have the capability to manage this event as it is run away from the project, so it cannot be controlled, therefore it takes a higher score.

It is important to define the competent person that can manage the event in case it does occur.

This method is based on probability and impact (consequence) and uses the numbers three, two, and one to rank what is high, medium and low, respectively. But for manageability, use the numbers one, two, and three to rank what is high, medium, and low, respectively.

In defining impact, the pessimistic one ranks everything as highly probable, while the carless one ranks everything as less probable. So, to control the meeting, be active and define the probability as follows:

- High probability – if the probability of occurrence is higher than 50%
- Medium probability – if the probability of occurrence is 10-50%
- Low probability – if the probability of occurrence is less than or equal to 10%

The facilitator and the team leader shall define the criteria before start assessing the risks. This criterion is different from one project to another based on size and budget. The assumed criteria for the example in Chapter 3, as will be discussed in Section 8.8, is as follows:

- High consequence – when the risk becomes reality and the impact on the project will cost higher than or equal to $500,000
- Medium consequence – when the risk becomes reality and the impact on the project will cost between $50K and $500K
- Lower consequence – when the risk becomes reality and the impact on the project will cost less than $50K.

8.6   Risk Response Planning and Strategies

The next task that must be done in our risk management system is risk response planning. At this stage, we have discovered all of the risks known to date and have an iterative process for discovering new risks as the project progresses. We have evaluated the risks and assessed their impact and probability of occurrence. We have prioritized the risks in their order of importance. We now must decide what to do about them. This is risk response planning.

Risk response planning is the process of developing the procedures and techniques to enhance opportunities and reduce threats to the project's objectives. In this process, it will be necessary to assign individuals who will be responsible for each risk and generate a response that can be used for each risk.

Risk response strategies are techniques that will be used to reduce the effect or probability of the identified, or even the unidentified, risks.

Of course in the case of opportunities, we should want to increase the probability and increase the impact. The opportunity can be exploited by adding resources to encourage and maximize the effect. Opportunities can be shared. In the case where our own organization is not able to maximize an opportunity, a partnership or other arrangement with another organization may be made where both organizations benefit in a greater way than one of them can. By enhancing an opportunity, we can maximize the drivers that positively impact the risks. Both impact drivers and probability drivers may be enhanced.

In terms of the risk strategy that should be employed, a qualitative or quantitative evaluation of the severity of the risk will be a guideline as to how much time, money, and effort should be spent on the strategy to limit the risk.

8.7   Risk Monitoring and Control

Risk monitoring and control is the process of keeping track of all the identified risks and identifying new risks, as their presence becomes known and residual risks that occur when the risk management plans are implemented on individual risks. The effectiveness of the risk management plan is evaluated on an ongoing basis throughout the project.

Figure 8.5 presents the steps of risk monitoring and control. After obtaining the risk priority for each risk, you should obtain a solution with the project team and then agree on the emergency plan and, according to every risk, define the monitoring system that you will follow as the project manager.

When a risk is apparently going to take place, put a contingency plan into place.

If there is no contingency plan, then the risk is dealt with on an ad hoc basis using what is termed a "workaround." A workaround is an unplanned response to a negative risk event. A corrective action is the act of performing the workaround or the contingency plan.

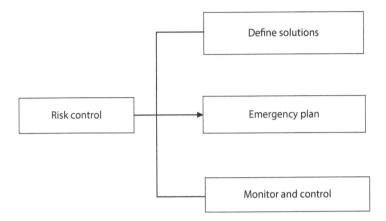

Figure 8.5 Steps in controlling the risks.

The concern of the project manager and the project team is that risk responses have been brought to bear on the risk as planned and that the risk response has been effective.

After they have observed the effectiveness of the risk response, additional risks may develop or additional responses may be necessary.

Risk management is a continuous process that takes place during the entire project from start to end. As the project progresses, the risks that have been identified are monitored and reassessed as the time that they can take place approaches. Early warning indicators are monitored to reassess the probability and impact of the risk. As the risk approaches, the risk strategies are reviewed for appropriateness and additional responses are planned.

Risk assessments, reviews, and audits may be performed periodically to review the probability and potential impact of risks that have been identified and are nearer to their possible occurrence. Risks that have already taken place can be reviewed and audited to assess the effectiveness of the risk response.

As each risk occurs and is dealt with or is avoided, these changes must be documented.

Good documentation ensures that risks of this type will be dealt with in a more effective way than before and that the next project manager will benefit from these lessons learned.

8.8   Example

Use the same example we discussed before in Chapter 4 about a small project for a new gas compressor.

Table 8.4 Risk Assessment Weight and Monitoring Sheet.

ID	Description	Owner	Probability H> = 50% = 3 M 10–50% = 2 L < = 10% =1	Overall consequence H = > 500K$ = 3 M = 50–500K$ = 2 L< = 50K$ = 1	Manageability L = 3 M = 2 H = 1	Score Max 27	$K if risk comes true	Weeks delay If risk comes true	Risk reduction action	Date Identify	Date Close
1	manufacturer delivery on time	SCM	1	3	2	6			To monitor performance regularly		
2	pump not match with specs requirement	ME	1	3	1	3			Review specs with the manufacture/ clarification meeting with the vendor		
3	anchor bolt and dimension problems during construction	CE	1	1	1	1			Review drawings with machine		
4	permits to work from operation	CM	1	1	1	1			a meeting with operation		

#	Description	Resp					Notes
5	Shut down plan coordination	CM	3	3	2	18	Meeting with operations
6	Missing bolts for thief	CM	1	1	2	2	security guard
7	weather effect during construction (fog)	CM	2	1	2	4	Late work hours procedure/ telecom.
8	H2S leakage in plant	CM	1	1	3	3	Procedure to overtime policy
9	Security pass to enter plant	CM/SD	2	2	2	8	Review security every month with the contractor
10	Procurement approval	SCM	1	2	2	4	Sign by hand/ presenta-tion to management
11	lake in labour language	CM	1	1	2	2	training matrix
12	Pump truck manoeuvre	HSE	1	2	1	2	Discuss moving by location layout

(Continued)

Table 8.4 (Cont.)

ID	Description	Owner	Probability H> = 50% = 3 M 10–50% = 2 L < = 10% =1	Overall consequence H = > 500K$ = 3 M = 50–500K$ = 2 L< = 50K$ = 1	Manageability L = 3 M = 2 H = 1	Score Max 27	$K if risk comes true	Weeks delay If risk comes true	Risk reduction action	Date Identify	Date Close
13	crane mobilization	CONT	2	3	2	12			Check with the crane		
14	late of concrete delivery	CONT	1	2	1	2			Confirm with the contractor		

Where: CM is the construction manager

SCM supply chain manager

CONT is the contractor

SD security department

HSE health, safety and environment department

ME mechanical engineering department

CE Civil engineering department

When looking at the project, everyone may jump to conclusions and define what he or she is afraid might occur during the project. In fact, when we go through the meeting using brainstorming, we may find the output from the following table.

8.9 Operations risk

In some old oil and gas companies, due to the mature facilities, their target project may switch to a maintenance project. If we make a comparison between Project and Maintenance Organization and Culture, it is clear that the concept of a project differs fundamentally from that of daily or routine operations and maintenance; it follows that a number of principles and conceptions of project management must also diverge from those followed in the realm of maintenance management.

In maintenance management, we tend, generally, to focus on maintaining the facilities' reliability during its lifetime production.

Project management may be defined as the planning, organization, direction, and control of all kinds of resources in a specific time duration for achieving a specific objective comprised of various financial and non-financial targets.

This should help clarify the difference in outlook of the project manager and the maintenance manager. The project manager's goal is to finish the project on time. Then, he evaluates where he will relocate after finishing the project. On the other hand, the maintenance manager never wants daily production to stop and cannot dream of work stopping as distinct from the project manager's goal of overall task completion.

The challenge here is in case of a major rehabilitation project, which is a major maintenance project or "brown field" project. So my question: Do the engineers in engineering on the construction phase in a brown field project have the same competency as the new, "green field" project?

Sure this is a big difference and an example of that. A big construction contractor performed very big projects in oil and gas, but when he obtained a maintenance contract, the company failed because the team and his organization is project oriented rather than maintenance oriented.

The project scope should be defined clearly and any deviations have a major control from the site. The project management technique is mainly different in new construction from the brown field. As in the brown field, the scope cannot be defined precisely, as when you do the construction scope, unforeseen circumstances may happen as you are working in mature facilities, so change of scope shall happen. Therefore, you should have a management system that overcomes the changes smooth and fast. On the other hand, the construction team should have an experience in maintenance and, practically, they do not have this experience. In most cases, the team can be easy to work with in construction, but to do the visa versa you will fail. Any engineer can work in new construction, but for brown filed or maintenance, it depends on experience.

8.10   Methods of Risk Avoidance

There are many ways to reduce the risk in general and these methods depend on eliminating the source of risk or transferring the risk to a third party. If this is done with a firm fixed-price contract, the risk is effectively transferred to the vendor.

Generally, in firm fixed-price contracts, the vendor will always raise the price of the service to compensate for the effect of the risk. In addition to that, the warrantees, performance bonds, and guarantees are additional methods for transferring risk.

The following techniques can be used to reduce and avoid risks:

- Clarifying requirements and objectives
- Improving communication
- Obtaining information
- Acquiring expertise
- Changing strategy
- Reducing scope
- Adopting familiar approach
- Using proven methods, tools, and techniques

You can transfer risk by transferring liability and ownership through the following:
Financial means:

- Insurance
- Performance bond, warranty

Contractual means:

- Renegotiate contract conditions
- Use subcontractor in parts of the project
- Joint ventures/ teaming
- Risk-sharing partnership with client

- Target-cost
- Note limit to which risk can be transferred
- Note that transfer usually involves a price tag

Quiz

1. Which of the following describes the BEST use of historical records?
   - Estimating, life cycle costing, and project planning
   - Risk management, estimating, and creating lessons learned
   - Project planning, estimating, and creating a status report
   - Estimating, risk management, and project planning

2. At which step of risk management does a determination of risk mitigation strategies take place?
   - Risk identification
   - Risk quantification
   - Risk response planning
   - Risk response control

3. By which of the following techniques can you calculate the risk assessment?
   - Arrow diagramming method
   - Network diagramming
   - Critical path method
   - Program evaluation and review technique

4. One of the risks your team has discovered is a high probability that the separator you are constructing will not perform safely under operation pressure. In order to handle this risk, you have chosen to test the separator materials and review design. This is an example of risk:
   - mitigation
   - avoidance
   - transference
   - acceptance

5. Which part of the risk management process uses data precision as an input?
   - Risk management
   - Qualitative risk analysis
   - Quantitative risk analysis
   - Risk response planning

References

Bourdaire J.M., R.J. Byramjee & R. Pattinson, 1985, "Reserve assessment under uncertainty - a new approach"; Oil & Gas Journal June 10, 135–140.

Cozzolino, J., 1977. A Simplified Utility Framework for the Analysis of Financial Risk, Economics and Evaluation Symposium of the Society of Petroleum Engineers, Dallas, Texas. Paper 6359.

J. Davidson Frame, Project management competence, J. Davidson Frame and Jossey-Bass Inc., 1999. John P. Kotter, Leading change, Harvard Business School Press, 1996.

Market P., Gustar, M., and Tikalsky, P. J. (1993) "Monte-Carlo Simulation Tool For Better Understanding of LRFD", J. of Struct., Div., ASCE, Vol. 119, No.5, May, pp. 1586–1599.

Michael W. Newell "Preparing for the Project Management Professional (PMP) Certification Exam", AMACO, 2005.

Nikolaos Plevris, Thansis C. Triantafi llou and Daniele Veneziano, (1995) "Reliability of RC Members Strengthened With CFRP Laminates", J. of Struct Div., ASCE, Vol. 121, No. 7, July, pp. 1037–1044. Project Management Institute Standards Committee. A Guide to the Project Management Body of Knowledge. Upper Darby, PA: Project Management Institute, 2000.

Project Management Institute Standards Committee. A Guide to the Project Management Body of Knowledge. Upper Darby, PA: Project Management Institute, 2004.

"The Modified Delphi Technique - A Rotational Modification," Journal of Vocational and Technical Education, Volume 15 Number 2, Spring 1999, web: VT-edu-JVTE-v15n2: of Delphi Technique developed by Olaf Helmer and Norman Dalkey.

Rashedi, M.R. (1984), "Studies on Reliability of Structural Systems," Ph.D. Thesis, Western Reserve University, Cleveland, Ohio. Ray Tricker (1997).

"ISO 9000 for Small Business," Butterworth-Heinemann, 1997.

Rescher(1998): Predicting the Future, State University of New York Press, 1998.

Richard Freeman, (1993) "Quality Assurance in Training and Education," Kogan Page, London.

Sackman, H. (1974), "Delphi Assessment: Expert Opinion, Forecasting and Group Process", R-1283-PR, April 1974.

Brown, Thomas, "An Experiment in Probabilistic Forecasting", R-944-ARPA, 1972 - the first RAND paper. Stephen Grey "Practical Risk assessment for Project Management", John Wiley & Sons, 1995.5

Turner, J. Rodney (1993). The handbook of project-based management. McGraw-Hill, London. "The Building Commissioning Guide," U.S. General Services Administration Public Buildings Service Office of the Chief Architect, April 2005.

Also of Interest

From the Same Author

Construction Management of Industrial Projects: A Modular Approach for Project Managers, by Mohamed A. El-Reedy, ISBN 9780470878163. The most comprehensive treatment of the processes involved in the project management and construction of industrial projects. With more regulations, risks, and costs accumulating for new industrial construction projects than ever before, this book guides the engineer or project manager through this maze of issues for an efficient and economically successful construction project. *NOW AVAILABLE!*

Check out these other titles from Scrivener Publishing

Seismic Loads, by Victor Lyatkher, ISBN 9781118946244. Combining mathematical and physical modeling, the author of this groundbreaking new volume explores the theories and applications of seismic loads and how to mitigate the risks of seismic activity in buildings and other structures. *NOW AVAILABLE!*

Hydraulic Modeling, by Victor Lyatkher and Alexander M. Proudovsky, ISBN 9781118946190 Combining mathematical and physical modeling, the authors of this groundbreaking new volume explore the theories and applications of hydraulic modeling, an important field of engineering that affects many industries, including energy, the process industries, manufacturing, and environmental science. *Publishing in spring 2016.*

Fundamentals of Biophysics, by Andrey B. Rubin, ISBN 9781118842454. The most up-to-date and thorough textbook on the fundamentals of biophysics, for the student, professor, or engineer. *NOW AVAILABLE!*

i-*Smooth Analysis: Theory and Applications,* by A. V. Kim, ISBN 9781118998366. A totally new direction in mathematics, this revolutionary new study introduces a new class of invariant derivatives of functions and establishes relations with other derivatives, such as the Sobolev generalized derivative and the generalized derivative of the distribution theory. *NOW AVAILABLE!*

Reverse Osmosis: Design, Processes, and Applications for Engineers 2nd *Edition*, by Jane Kucera, ISBN 9781118639740. This is the most comprehensive and up-to-date coverage of the "green" process of reverse osmosis in industrial applications, completely updated in this new edition to cover all of the processes and equipment necessary to design, operate, and troubleshoot reverse osmosis systems. *NOW AVAILABLE!*

Pavement Asset Management, by Ralph Haas and W. Ronald Hudson, with Lynne Cowe Falls, ISBN 9781119038702. Written by the founders of the subject, this is the single must-have volume ever published on pavement asset management. *NOW AVAILABLE!*

Open Ended Problems: A Future Chemical Engineering Approach, by J. Patrick Abulencia and Louis Theodore, ISBN 9781118946046. Although the primary market is chemical engineers, the book covers all engineering areas so those from all disciplines will find this book useful. *NOW AVAILABLE!*